Liangzhu Culture

The Liangzhu Culture (3,300–2,300 BC) represented the peak of prehistoric cultural and social development in the Yangtze Delta. With a wide sphere of influence centred near present-day Hangzhou City, the Liangzhu City is considered one of the earliest urban centres in prehistoric China. Although it remains a mystery for many in the West, Liangzhu is well known in China for its fine jade-crafting industry; its enormous, well-structured earthen palatial compound and recently discovered hydraulic system; and its far-flung impact on contemporary and succeeding cultures. The archaeological ruins of Liangzhu City were added to the UNESCO World Cultural Heritage List in July 2019.

Liangzhu Culture contextualises Liangzhu in broad socio-economic and cultural backgrounds and provides new, first-hand data to help explain the development and structure of this early urban centre. Among its many insights, the volume reveals how elites used jade as a means of acquiring social power, and how Liangzhu and its centre stand in comparison to other prehistoric urban centres in the world.

This book, the first of its kind published in the English language, will be a useful guide to students at all levels interested in the material culture and social structures of prehistoric China and beyond.

Liu Bin* is Professor and Director of Zhejiang Provincial Institute of Cultural Relics and Archaeology. He has joined or directed excavations at the Fanshan, Yaoshan, Huiguanshan, and Nanhebang sites, and the Liangzhu City since 1985. His main research interests include the prehistory of the Lower Yangtze River and the archaeology of jade.

Qin Ling is Associate Professor of Neolithic Archaeology and Archaeobotany at the School of Archaeology and Museology, Peking University, Beijing. Her research interests include scientific research on Neolithic jades in Eastern China, early agricultural developments in the Lower Yangtze River and Southwest China, and comparative perspective on civilisational discourses across East Asia.

Zhuang Yijie is Associate Professor in Chinese Archaeology at the Institute of Archaeology, University College London. He applies geoarchaeological approaches to reconstruct ecologies of early agriculture and long-term land use changes in East, South, and Southeast Asia. He is also interested in the comparison of diverse trajectories to social complexity in these regions.

* The Chinese names in this book all start with their surnames, followed by their given names.

Liangzhu Culture

Society, Belief, and Art in
Neolithic China

**Edited by Liu Bin, Qin Ling,
and Zhuang Yijie**

LONDON AND NEW YORK

First published 2020
by Routledge
2 Park Square, Milton Park, Abingdon, Oxon, OX14 4RN

and by Routledge
605 Third Avenue, New York, NY 10017

First issued in paperback 2021

Routledge is an imprint of the Taylor & Francis Group, an informa business

Publisher's Note
The publisher has gone to great lengths to ensure the quality of this reprint but points out that some
imperfections in the original copies may be apparent.

British Library Cataloguing-in-Publication Data
A catalogue record for this book is available from the British Library

Library of Congress Cataloging-in-Publication Data
A catalog record has been requested for this book

Typeset in Sabon
by Newgen Publishing UK

ISBN 13: 978-1-03-208483-1 (pbk)
ISBN 13: 978-1-138-55740-6 (hbk)

Contents

List of illustrations

Figures

Tables

Contributors

Chen Minghui* 陈明辉 (Research associate and head of the Liangzhu Archaeological Workstation, Zhejiang Provincial Institute of Cultural Relics and Archaeology)

Fang Xiangming 方向明 (Professor and deputy director of the Zhejiang Provincial Institute of Cultural Relics and Archaeology)

Liu Bin 刘斌 (Professor and director of the Zhejiang Provincial Institute of Cultural Relics and Archaeology)

Qin Ling 秦岭 (Associate Professor, School of Archaeology and Museology, Peking University)

Sun Qingwei 孙庆伟 (Professor and director of School of Archaeology and Museology, Peking University)

Wang Ningyuan 王宁远 (Professor and director of the Liangzhu Hydraulic System research project, Zhejiang Provincial Institute of Cultural Relics and Archaeology)

Zhao Hui 赵辉 (Professor, School of Archaeology and Museology, Peking University)

Zhuang Yijie 庄奕杰 (Associate Professor, Institute of Archaeology, University College London)

* The Chinese names in this book all start with their surnames, followed by their given names.

Acknowledgements

We are indebted to the following institutions for supporting the exhibition entitled "*Power and Belief* 权力与信仰" in 2015 at Peking University, Beijing, and for supporting us to expand some contents of the same-titled Chinese exhibition catalogue into this book. They are:

- School of Archaeology and Museology, Peking University, Beijing
- Zhejiang Provincial Institute of Cultural Relics and Archaeology, Hangzhou
- Liangzhu Museum, Liangzhu, Hangzhou
- Yuhang Museum, Yuhang, Hangzhou

We thank National Education Ministry of China's grant for funding our project on "Technology and Civilization: The Foundation of Early Chinese Civilizations in a Perspective of Jade-stone Handicrafts" (project number: 16JJD780004).

We thank Professor Xu Hong and Erlitou Archaeology Team, Institute of Archaeology, CASS (Chinese Academy of Social Sciences) and Professor Meng Huaping, Hubei Provincial Institute of Cultural Relics and Archaeology for providing the photos for Chapter 6.

We especially thank Mr. Wu Zhenglong for photographing most of the Liangzhu jade for this book, as well as for designing the book cover.

Preface

In cooperation with the international conference, 'Dialogue of Civilizations: Comparative Studies on Early Civilizations in a Global Perspective', co-hosted by Peking University and University College London, Zhejiang Bureau of Cultural Heritage and Yuhang District Government organised a grand exhibition entitled 'Power in Things: New Perspective on Liangzhu' at the Arthur M. Sackler Museum of Art and Archaeology, Peking University. To promote the public understanding of the Liangzhu Culture, several organisations, including Zhejiang Provincial Institute of Cultural Relics and Archaeology, School of Archaeology and Museology of Peking University, Centre for Chinese Archaeological Study of Peking University, Liangzhu Museum, and Yuhang Museum of Hangzhou jointly compiled and published the catalogue *Power in Things* for this exhibition.

Since as early as 1936 when Mr. Shi Xingeng discovered black pottery and stone objects at Liangzhu, the excavation and research of Liangzhu Culture have gone through 80 years. However, it was not until 1973 when archaeologists found some elite Liangzhu tombs buried with *cong* tubes and *bi* disks that we began to realize that these ancient jade, originally thought to be Zhou or Han period, was indeed made by the Liangzhu people 5,000 years ago. Then, in 1986, 1987, and 1991, local archaeologists found more and more large burials within the Liangzhu site cluster, such as Fanshan, Yaoshan, and Huiguanshan. Surprised at the stunning beauty of this ancient jade, scholars started to explore the social structure and hierarchical system of the Liangzhu Culture. Now, after accumulating a great deal of material and good understanding of these Liangzhu archaeological remains, we are fully confident to say that Liangzhu was an early complex society, entering the stage of early state.

In 1986, archaeologists identified a completely structured 'shenhui insignia' on a big *cong* tube in tomb no. M12 at the Fanshan cemetery. The 'shenhui insignia' is the main decorative pattern combined with the God of nature and the God of ancestors and has recurred on Liangzhu jade items. Whether it was represented in detailed or simplified forms, the use of this stylistic pattern was highly unified across a large area. This decorative pattern was the materialisation of the shared belief system within the whole Liangzhu Culture area. The 'insignia' might represent a mighty warrior who was remembered by the Liangzhu community by joining the many different tribes along the Tai Lake area in the Lower Yangtze River, thus established his far-reaching cult of personality. Consequently, he was remembered by the Liangzhu people generation after generation and his image was incorporated and represented in the jade decorative system. This is likely the internal connotation of Liangzhu jade, which is not only part of the decorative scheme but also the material carrier of Liangzhu kinship, military power, and religious authority.

With the discovery of the Liangzhu City in 2007, the archaeology of Liangzhu has made many breakthroughs within the past ten years. Through coring and excavation within an area of 100 km², archaeologists have reached a comprehensive understanding of the main part of the whole site complex and have sketched out the general picture of the ancient city. The Liangzhu City is mainly composed of a core area, the ceremonious sites outside the city, and the hydraulic system. The core area is in a 'centripetal' structure enclosed by three layers of walls. The Mojiaoshan platform is located right in the centre of the enclosed area, occupying about 30 ha. The inner city is surrounded by a 6 km wall and occupies an area of about 300 ha, while the outer city sized about 800 ha in total. Ten kilometres away from the ancient city in the northern and north-western direction, is a large-scale hydraulic system. Such a huge construction project could only be accomplished with the existence of a highly centralized social power and efficient managerial organisation. After 5,000 years, the housing structures of Liangzhu period have long gone, we cannot see the Mojiaoshan palaces on the platform, but this does not impede our perception of the Liangzhu City through its incredibly rich material remains. When one stands on the Mojiaoshan platform, looking around, to the north, west, and south, in all three directions are equidistant mountains, with the Liangzhu City situated within the middle of 'heaven and earth'.

The Liangzhu Culture gradually declined around 4,300 years ago and the main city area failed to maintain its position as a power centre in later history. However, the innovation of Liangzhu civilisation has deeply rooted in societies of later Chinese history. For instance, the 'centripetal' structure remained the major practice of urban planning throughout later history; the jade forms such as *bi* disks and *cong* tubes remained important sacrificial items even in Zhou period (ca. 1050–221 BC) and were documented extensively in transmitted texts such as *Rites of Zhou*. Since then, *bi* disks and *cong* tubes have been the representatives of high-class jade objects and carry very important ritual meanings. The concrete meaning of the Liangzhu '*shenhui* insignia' pattern had become blurred even before the fall of Liangzhu, but the ritual belief and worship for ancestors remained and become a consistent cultural tradition both in ancient China and today.

Catherine Xinxin Yu, Wang Shaohan, and Tang Xiaojia translated the chapters. Dr Charlene Murphy proofread the whole draft. Chen Minghui also helped with the editorial work.

丁沙地 Dingshadi

高城墩 Gaochengdun

寺墩 Sidun

昝庙 Zanmiao

嘉陵荡 Jialingdang

邱承墩 Qiuchengdur

堰南 Yannan

越城 Yuecheng

太湖 *Taihu Lake*

龙南 Longna

邱城 Qiucheng

杨家埠 Yangjiabu

钱山漾 Qianshanyang

塔地 Tadi

新地里 Xindili

芝里 Zhili

喇叭浜 Lababang

辉山 Huishan 玉架山 荷叶地 Heyedi
             Yujiashan
                     盛家埭
横山 Hengshan   Shengjiadai
             茅山 Maoshan
             后头山 Houtoushan
良渚遗址群   徐步桥
Liangzhu Site Cluster   Xubuqia

杭州 *Hangzhou*

古荡 Gudang

大麦山 Damaitu

小青龙 Xiaoqinglong

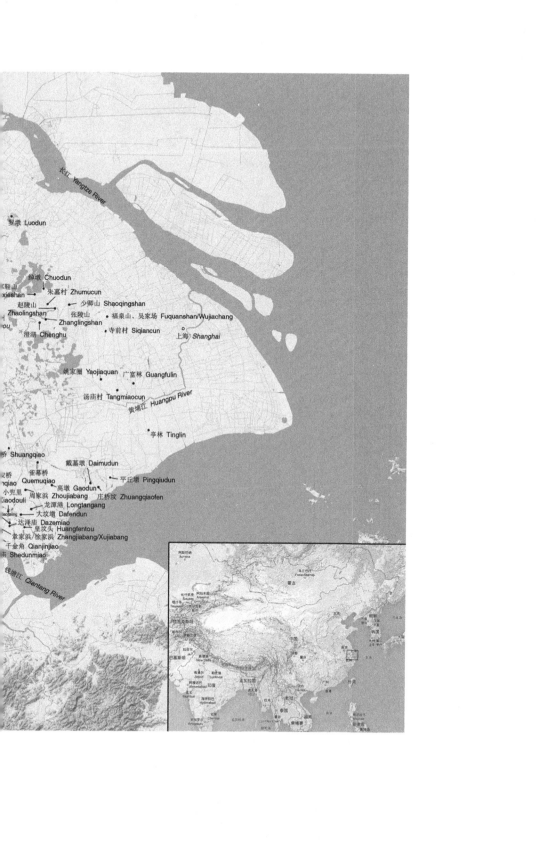

长江 Yangtze River

罗墩 Luodun

绰墩 Chuodun
(鞋山)　朱墓村 Zhumucun
xieshan
赵陵山　少卿山 Shaoqingshan
Zhaolingshan
张陵山　福泉山、吴家场 Fuquanshan/Wujiachang
Zhanglingshan
ou
寺前村 Siqiancun
澄湖 Chenghu
上海 Shanghai

姚家圈 Yaojiaquan　广富林 Guangfulin

汤庙村 Tangmiaocun　黄埔江 Huangpu River

亭林 Tinglin

桥 Shuangqiao
雀幕桥　戴墓墩 Daimudun
反桥
nqiao　Quemuqiao　高墩 Gaodun　平丘墩 Pingqiudun
小兜里　周家浜 Zhoujiabang　庄桥坟 Zhuangqiaofen
iaodouli
aojiang　龙潭港 Longtangang
大坟墩 Dafendun
达泽庙 Dazemiao
皇坟头 Huangfentou
章家浜／徐家浜 Zhangjiabang/Xujiabang
千金角 Qianjinjiao
庙 Shedunmiao
钱塘江 Qiantang River

1 Situating the Liangzhu Culture in late Neolithic China

An introduction

Liu Bin, Qin Ling, and Zhuang Yijie

The volume you are reading is one of the first English volumes dedicated to a very important subject, that of the Neolithic culture from the Lower Yangtze River, one that is, unfortunately, not well known among non-Chinese readers. The second to sixth chapters of this volume were written by eminent Chinese archaeologists, on the frontline of excavations and research on the Liangzhu Culture, for the catalogue of a special exhibition held at the Arthur M. Sackler Museum of Art and Archaeology, Peking University, March–April 2015. The original catalogue won the 'Top Ten National Archaeology and History Books of the Year' award in 2016. Necessary introductions and explanations of unfamiliar background information and difficult terminologies have been included in the chapters of this volume to assist English readers. This introductory chapter continues this purpose and contextualises the key issues discussed in the following chapters within the broader sociopolitical backgrounds of late prehistoric China. Combined with the other five chapters, this introductory chapter will further help readers navigate in the ever-expanding scope of Chinese archaeological material.

1.1 What is the Liangzhu Culture?

1.1.1 Short research history of Liangzhu

The Liangzhu Culture (3300–2300 BC) represents one of the classic examples in which the archaeological research of a highly developed regional culture can contribute to our understanding of the origins of Chinese civilisations. The Liangzhu site at Liangzhu Town, Hangzhou, was first found and excavated by Shi Xingeng in 1936. Initially it was considered a local type of the Shandong Longshan Culture in the Lower Yangtze River due to its shiny black pottery resembling that of the Longshan Culture (Shi 1938). Archaeological research on the Liangzhu Culture and its associated temporal and spatial contexts has undergone three major stages. Between 1936 and 1980, archaeological research was focused on the establishment of pottery typology, which was popularly used to reconstruct cultural lineage and relative chronology by most Chinese archaeologists. It was becoming clear that the Liangzhu Culture was earlier than the Longshan Culture in north China. The late 1970s to 1980s was a period when a major breakthrough of the Liangzhu archaeology was made. This began with the discovery and excavation at the Caoxieshan cemetery in Jiangsu, followed by the excavations of a series of the rich burials from cemeteries at Sidun in Jiangsu, Fuquanshan, in Shanghai, and Fanshan, Yaoshan, and Huiguanshan in Zhejiang. These exceptionally well-preserved burials (unfortunately, most organic matter and human skeletons were destroyed due to the acidic soil conditions) provide a rare glimpse into the life and death of the Liangzhu elite circle, which was centred on the production, consumption, and circulation of jade items with meticulously carved motifs. These discoveries draw the curtain for more research on the Liangzhu jade. Liangzhu jade has become the main indicator of social status, gender division, ritual and religious practices, and development of social complexity. These issues have dominated Liangzhu archaeology since the 1980s. Because Liangzhu jade is so distinctive in its raw material, shapes, and carved symbols and so widely influential among Chinese Neolithic jade, it is regarded as representative of the most complicated craftsmanship in prehistoric China. From 2007, the discovery of the city walls has pushed the archaeological investigation of the Liangzhu society to a new level. Archaeologists have now realised that the enclosed area and its surroundings at Liangzhu were an early urban centre, the so-called Liangzhu City. This revised understanding of the complexity of Liangzhu society reached its climax with the recent discovery of the Liangzhu hydraulic system (see Chapter 2; Qin 2013). This enormous infrastructure project has revealed the truly unprecedented scale of landscape transformation that the Liangzhu society had achieved, which could not have been possible without a state-level managerial system.

1.1.2 Distribution of the Liangzhu Culture

The accumulation of rich archaeological material in the past 80 years has enabled us to have a relatively complete understanding of many aspects of Liangzhu society. These include the chronology, the life and death of both the elite groups and normal villagers, production and consumption patterns of elaborate objects and economic goods, the relationship between its centre and peripheries, its wider influences on other contemporary regional archaeological cultures, and its legacy to later Neolithic and historical cultures. The Liangzhu pottery assemblage is characterised by its black pottery with shining surfaces and elaborate motifs that are similar to those seen on jade and lacquer objects. The most common pottery set includes the *ding* tripod vessels, the *dou* stemmed dish, the *shuangerhu* two-eared bottles, and *guan* jar (Figure 1.1). The pottery assemblages among the contemporary archaeological

Figure 1.1 Typical pottery of the Liangzhu Culture from Bianjiashan (top left: *ding* tripod; top right: *dou* wide-stemmed dish; bottom left: *hu* two-eared bottle; bottom right: jar)

sites in the region are distinctively similar in terms of pottery types, design, and production technology. Based on this pottery typology, more than 600 Liangzhu Culture sites have been found around the Lower Yangtze area. The most densely populated area was found at and surrounding the Liangzhu City. Around 300 sites have been found with Liangzhu deposits within 300 ha inside and outside of the walled city. According to our estimation of population density of c. 100–150 person per km² at the site of Xiantanmiao, 22,900–34,350 people would have lived in this area. This calculation excludes the palatial areas as it is based on data derived from a small settlement site functioning as a basic economic production unit. The spot population density at the palatial areas such as the Mojiaoshan platform would have been higher (cf. Liu 2006). Several site clusters have been recognised outside this core zone (see Figure 4.25; Nakamura 2003). The Linping Site Cluster was around 20 km to the northeast of the Liangzhu City. Around 20 sites have been found to date which form a fully functional sub-centre, including cemetery, residential, and production sites (Zhao 2012). The well-preserved paddy fields located in the transitional area between the foothill and the alluvial plain at Maoshan, not far away from the cemetery on the foothill, suggest that this sub-centre would have been economically self-sufficient. However, excavations at the Yujiashan and Hengshan cemeteries revealed that the elites buried here also had access to some of the high-quality jade objects, pointing to a shared symbolic system or ideology between the Linping sub-centre and the Liangzhu centre. Similar ritual activities surrounding jade were practiced by these local elites in their imitation of the elite circle at the Liangzhu centre. The Linping case suggests that while the sub-centre might be economically self-sustained, it was still tightly connected to the core area through jade and the symbolic system associated with it. Further out of the core area, there were more site clusters concentrated around the Taihu Lake region (Qin 2013). These clusters were used in different phases of the Liangzhu Culture period and had rich burials containing high-quality jade objects similar as those from the Liangzhu centre.

Beyond this core zone were the regional archaeological cultures that had either very strong links with the Liangzhu core zone or were 'colonised' by Liangzhu people. The Liangzhu influence reached as far north as northern Jiangsu Province, to the west to what is nowadays Nanjing City, and as far south as southern Zhejiang Province (Figure 1.1). Archaeological sites in these areas contained jade objects that were either directly 'imported' from the Liangzhu centre or imitation of the typical Liangzhu jade. Thus, the ceramic assemblages would contain both Liangzhu-style ceramics and local-style ceramics. The most interesting case comes from the excavation at the Huating site (Nanjing Museum 2003), close to the Shandong Dawenkou Culture area. This cemetery consists of both southern and northern parts. The southern part of the cemetery was used first. The burials contain predominantly Dawenkou-style ceramics over the course of the whole occupational period. Artefact burial assemblages in the northern part of the cemetery contained different cultural influences of the Dawenkou Culture, the Xuejiagang Culture, and the Liangzhu Culture. Of particular significance were the jade items in these burials: Some of them (e.g., Liangzhu-style *cong* tubes) were direct from Liangzhu Culture, while others were local products with Liangzhu, Dawenkou and other influences (Nanjing Museum 2003; Huang 2011). This highly mixed archaeological assemblage clearly indicates long-term interactions between the Liangzhu and contemporary cultures in the north, even though the Liangzhu influence here was not as dominant as it was to its immediate neighbours.

1.1.3 Landscape of the Liangzhu City

The Liangzhu Culture is representative of Neolithic China not only because of its highly developed material culture but also because of the level of landscape transformation. This type of engineered landscape is characterised by large-scale earthen constructions, completely different from the stone-based architecture in prehistoric Europe and the Near East (Jarzombek 2014). The structure of the Liangzhu City and its enormous scale are discussed in Chapter 2. Several points are worth reiterating and expanded here within a broader context.

First, much endeavour in prehistoric archaeology of China has been spent on the excavation and research of walled sites or cities. Many prehistoric cities, the majority dated to between 3000 and 2000 BC, have been discovered through this unprecedented archaeological campaign (Zhao and Wei 2002). However, 'cities' as a definition remain poorly defined. Can we really call these walled sites cities? Were they structurally and functionally different or similar to modern cities (in an economic, political, or cultural sense)? The Liangzhu City is one of the rare cases through which we could visualise and test some of the propositions of Chinese prehistoric cities made by archaeologists. The structure of the Liangzhu City is characterised by one circle of inner earthen walls and several discontinuous earthen-mound settlements encircling the outer city range and an enormous artificial platform, Mojiaoshan, in the centre as well as many other cemeteries and smaller mounds. Details of the structure of the Mojiashan platform are given in Chapter 2. The Liangzhu elites lived on this palatial complex (30 ha), 6–12 m higher than the surroundings, with palace foundations, storage pits, and many other features. The elites, their life and death, have been separated forever from those who might be regarded as 'commoners'. Such a separation is not unique among the late Neolithic walled sites and as Liu et al. argue in Chapter 2, it signalled a fundamental departure from the spatial structure of farming villages dominated by small houses without a clear centre. The elites living on such high, central mounds could oversee the surrounding landscapes; being visible in the centre would have been an important way to display their power. These characteristics are comparable to political centres of proto-historical and historical periods, where elites and the ruling classes also occupied the centre of these cities.

Second, but unlike other prehistoric 'cities' of China, where our understanding of their wider economic and environmental contexts remains ambiguous, as detailed in Chapter 2, excavations have convincingly illustrated that the landscape of the Liangzhu City and its hinterland was to a large extent an anthropogenic one with clearly defined functional divisions. Both the scale of landscape transformation, characterised by the well-organised earthen works, and the density of population were unprecedented at this time and what we could call an early form of urban centres in prehistoric China. Of particular importance was its well-connected transportation system. In this highly engineered landscape criss-crossed by both natural and artificial water bodies, around 51 artificial canals (measuring c. 30 km in total) have been found according to recent surveys (see Chapter 2). These were connected to the eight so-called water gates of the Liangzhu City and, presumably, also linked to the hydraulic system located to the north and north-east of the city. Wang et al. (2019) have suggested that one of the functions of this water system was to transport stones for the construction of the walls from the mountainous stone-collection sites to its north. In the same vein, it may be envisioned that this system was used to transport other material goods in and out of the city. This transportation advantage should not be overlooked. Not only was it vital in terms of the logistics of the construction of the urban centre, it would have also played a key role for the elites to exploit the economic resources from the surrounding areas. This is very similar to the city of Uruk and its relationship with its neighbours in Mesopotamia. We will return to this point later.

1.2 Acquisition of social power in late Neolithic China

1.2.1 Some historical tendencies of the late Neolithic China

The late Neolithic period in China experienced profound socio-economic changes and technological advancements that led to the formation of early complex societies or states. However, a central China perspective remained predominant in the scholarly investigation of the trajectory of state formation for a long time, hindering a more robust understanding of the centre-periphery interactions and their importance to state formation processes. Thanks to the recent surge of archaeological discoveries in different regions of China, we have now had a better appreciation of the following aspects of late-Neolithic societies in China: (1) Technological advancements in these different regions did not happen simultaneously; (2) those regions previously regarded as marginal in terms of their socio-economic structures and technological levels were in fact often quite advanced on certain aspects; and (3) mechanisms and stimuli that drove economic and technological developments differed greatly among these different regions (Ministry of Science and Technology 2009; Han 2015). A multicentre perspective is currently taking shape, calling for the re-evaluation of the roles of these regions in the origins of Chinese civilisation, a process that is characterised by increasing and intensifying regional interactions between these regions and beyond. The flow of goods as well as technologies helped to shape a powerful force, like a spinning whorl as described by some scholars (cf. Zhang 2017), that not only drew societies located in different geographic regions closer together but also made the different sectors of the society more intricately intertwined (Liu 2003). Despite these regional differences, there are also several general tendencies that can be observed in these regional archaeological cultures.

First, the beginning of the fourth millennium BC saw dramatic changes in regional settlement patterns in different regions, with pronounced changes in the quantities of settlements and the appearance of very large walled sites (Liu and Chen 2012; He 2013; Zhao, C. Q. 2013), indicative of profound demographic changes and/or social reorganisation (cf. Wagner et al. 2013). These large-sized walled sites became regional centres and had increasingly well-defined functional divisions within and/or outside the walled areas. Two important examples are the Shijiahe Walled Site in the Middle Yangtze River and the Tonglin Walled Site in Shandong. The size of the former reached around 120 ha. Inside and around the walled areas were remains of intensive economic and ritual activities (Zhang C. 2013). Measuring around 230 ha, the Tonglin site is composed of nine functionally different sections, with the central one considered a city enclosed by rammed earth walls (Sun 2013). This central walled area was used over multiple periods and by the late phase reaching around 35 ha in size.

Second, in terms of economic production, there is a general trend towards specialisation and intensification, supported by technological advancements, craftsmanship, and organisation of production (Flat 2016), and most importantly, the increasing demands from the elites and other sectors of societies due to various reasons (Liu 2003). The richest Middle Taosi period tomb (no. IIM22) discovered at the Taosi Walled Site, for instance, contains at least 118 surviving grave goods in the niches and burial chamber, including many ceremonial jade, elaborate painted pottery vessels with vivid designs, lacquer wares, and many other items, as well as a handful of full or half pig skeletons (He 2009, but He 2013 states the number was 78). The tomb occupant is thought to be a king, who appears to have unlimited access to all kinds of resources, including locally produced products and long-distance exchanged items. Similar examples can be seen in many other cemeteries (e.g., the Longshan Culture Zhufeng cemetery in Shandong) (Luan 2013). These elite burials suggest a close link between elite consumption of luxury goods and technological development of production, a connection that is also prominent in the Liangzhu Culture as the chapters in this volume will demonstrate. However, there are alternative models. As Hung has found in her detailed investigation of the Majiayao Culture pottery production and consumption, this society focused on limited types of painted pottery vessels and was more concerned with the quantity. While this concern drove the production to a very high level, the quality dropped significantly. In this society, we do not see the parallel development of intensification and quality improvement of craft production (Hung 2011).

Third, there is growing evidence of social stratification at these regional centres. The material evidence for this mainly comes from excavations of cemetery sites, in which the quality and quantity of

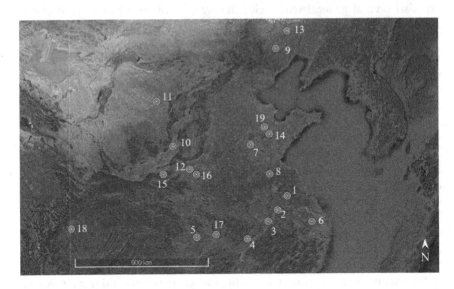

Figure 1.2 Locations of key sites mentioned in the chapters: 1. Longqiuzhuang;
2. Beiyinyangying; 3. Lingjiatan; 4. Xuejiagang; 5. Shijiahe; 6. Songze;
7. Dawenkou; 8. Huating; 9. Niuheliang; 10. Taosi; 11. Shimao; 12.
Erlitou; 13. Xinglongwa; 14. Zhufeng; 15. Xipo; 16. Wangchenggang;
17. Panlongcheng; 18. Sanxingdui; 19. Tonglin

burial goods vary significantly within and between cemeteries. Some rich
burials would contain hundreds of elaborate artefacts, including jade,
lacquer, and, later on, bronze items that were not produced locally, while
poor burials would only be accompanied by several pots at the maximum. Even though meanings of such differences vary as depending on
the local institutions, they might reflect social, wealth, or other status
markers of the buried person. These material differences point to radical
and unprecedented changes within these late Neolithic societies across
China (Figure 1.2).

1.2.2 Debates on the origins of Chinese civilisations

The debate on whether the Early Bronze Age site Erlitou should be considered
representative of an early civilisation seems to have come to a dead end,
due to the lack of a mature writing system at Erlitou, which is regarded,
by some scholars, to be the prerequisite of a civilisation. However, Erlitou
and many of its predecessors in the Longshan period have bred some of the
key criteria for 'civilisation'. As Thorp (2006, p. 57) accurately summarises

> for most Chinese scholars, the Erlitou Culture represents the birth of
> civilization in the heartland of China Proper. This archaeological culture

produced a major site with key attributes of a complex society: large-scale architecture, bronze metallurgy, a stratified social order, and elaborate rituals. Only writing is missing from a normative definition of civilization.

To us, many of these debates miss an important point, that is the investigation of how these traits developed and became the distinctive characteristics of early complex societies on the eve of civilisation.

Su Bingqi was among the early scholars who systematically articulated theories for early state formation in China. In his well-known 'three stages' model, he proposed that ancient Chinese civilisation underwent a three-step development from cultures to cities to states (Su 2000) and that there were primary, secondary, and continuous types of early states across different cultural regions of China. The Hongshan Culture in the Liao River valley of north-east China, with its goddess temple, altars and rich burials, and hierarchical settlement patterns, was an ancient local state higher than a tribal polity (Su 2000; Li 2016). For its emphasis on the importance of regional cultures such as the Hongshan Culture to the emergence of civilisation, Su's theory stimulated a passionate search of new archaeological evidence in these local regions. While acknowledging the importance of the Hongshan Culture, Yan Wenming thinks that many Longshan societies in the Yellow River valley were close to the chiefdoms defined in Service and Earle's works (Yan 1997). Adopting this evolutionary perspective, Li Boqian (2016) and others (e.g., Zhang 1995) suggest that Liangzhu represented a kingdom in south China, a developmental stage succeeding the Hongshan Culture and its contemporaries. Taosi is considered Liangzhu's counterpart in north China. Among many features appearing at both Liangzhu and Taosi, such as large-scale earthen works and evidence of social stratification, kingship and thearchy were the two most important features that fundamentally defined the Liangzhu and Taosi kingdoms. Indeed, according to many scholars, this dichotomy between kingship and thearchy differentiates Taosi in the north from Liangzhu in the south. Undoubtedly, while these theories significantly contributed to the nuanced understanding of the multiple origins of Chinese civilisations, they tend to focus on big narratives. There is a need to examine the complex models of social development under dramatically different social, economic, and environmental conditions. Especially, we need to avoid taking a unilinear perspective, which assumes that cultures or societies always experience steady growth, from simple forms of social organisations to complex societies. Instead, complex societies with highly stratified organisations might undergo radical reorganisation over a short period and their subsistence strategies and social structures would change too, not necessarily in a more complex direction, but one that may best accommodate new scenarios. In the ecologically sensitive or culturally flexible zones, the shift in economic and cultural practices might be necessarily linked to social or political changes. The late Neolithic communities living in the intermediate zone between the loess area and the Eurasian steppes, for instance, shifted back and forth between farming and pastoralism, depending primarily on climatic conditions (Frachetti et al. 2012; Spengler et al. 2014).

1.2.3 Multiple paths to social power

Having agreed that many late Neolithic societies could be considered early states or state-level political entities, at the centre of the scholarly debates on early state formation is how social power was acquired. Mann's influential volumes look at societies as 'organized power networks' and propose four main avenues through which power rose or is gained: Ideological, economic, military, and political, by mainly referring to Mesopotamia and European cases (Mann 2012). Many of these aspects have been dealt with or touched upon by the aforementioned models. For the proponents of the kingship-thearchy model, the social power of privileged individuals was achieved mainly through military actions and priesthood. Zhang Zhongpei, a close follower of Su Bingqi and his theory, thought that the tomb occupants of the rich Liangzhu burials formed a new social class above ordinary people (Zhang 2012). This protagonisation of powerful individuals, as informative and stimulating as it, is not without problems. First, these models are predominantly a manifestation of social structure based on mortuary data. Little effort is spent in integrating the mortuary evidence with other lines of evidence and we know very little how these privileged individuals exerted their power in daily life. More importantly, there is a pronounced lack of evidence for inter- and intraregional conflicts and/or violence at most regions (but cf. He 2013 for some evidence at Taosi). This is at odds with the commonly agreed idea among Chinese archaeologists that the *yue* axe is the material symbol of military power in late prehistoric China, whose prestigious status was derived from its function in battle or warfare. Such a relationship between axe and the military prowess, which then was translated into social power, was also prominent in Liangzhu society.

However, the Liangzhu case has further insights to offer to deepen our understanding of the formation of social power in early China. Li, Zhang, and others' classical narratives discussed previously predominantly focus on the acquisition of male power through military and ritual activities. Even though human skeletons are generally not well preserved at Liangzhu cemeteries, due to the acidic soil conditions, a close association between gender and jade has been established by some studies. Drawing on the rich mortuary data from the Liangzhu elite cemeteries, which offer a rare glimpse into elites' lives and power in Liangzhu society, Chapter 3 not only provides a detailed account of how the power of elite males (including a king) is realised through jade but also illustrates the significance of the jade objects that have a specific association with female elites. However, Qin also demonstrates how the social status of female elites was in decline in a society that was becoming increasingly patrilineal.

Her chapter also deals with another important aspect of power in Liangzhu society, that is 'power and space'. Not only does she notice the spatial difference between the living and the dead in the selection on burial objects and their display in tombs, she also cross-examines power and the changing spatial arrangements in the Liangzhu Site Cluster. As she puts it, '[T]he flux of spatial arrangements is also a process through which the Liangzhu Site Cluster established, consolidated and expanded its central position towards its neighbours.' It is also important to note that the agents behind this process should include the Liangzhu elites and non-elite members at the centre who participated in the planning and construction of the complex.

The last aspect, which is not dealt with by Mann and other scholars, but remains a salient characteristic at Liangzhu is the relationship between technology and power. The sophistication of the Liangzhu jade craftsmanship, as repeatedly discussed, is unparalleled among the Neolithic jade cultures in China (Rawson 1995; Qin 2013). This craftsmanship is clearly controlled within the limited circle of the Liangzhu centre elites who are both the producers and consumers of the jade. The tight association between jade products and jade crafts possibly co-evolved, as the Liangzhu elites continued to reinforce their power over control of resources and other aspects of jade and related industries and expand their influence. This technological knowledge was confided to the top elite circle and passed on to select generations.

In short, Liangzhu society represents a unique yet typical pathway towards social complexity in prehistoric China. The acquisition of social power also functioned as a nexus in which technology, monopoly of resources, performance, and other associated social and economic activities converged and interacted. The confluence of many of these powerful forces can be seen in other contemporary or later cultures in China (e.g., Post-Shijiahe and Sanxingdui) (Sun 2013; Zhang, C. 2013), but Liangzhu society remains the most prominent case to examine this power acquisition process with its enormous urban centre, well-regulated funerary hierarchy, jade system, and unprecedented influence in its region.

1.3 Liangzhu in comparison

The economic structure and organisation, the population level and labour investment, the trajectory to social power, and the interactive sphere of Liangzhu are comparable to those of other early urban centres in the prehistoric world. Needless to say, the environmental contexts and

socio-economic institutions of these early complex societies are dramatically different from each other. Few would disagree that a brief, but comparative discussion between Liangzhu and other contemporary urban
centres will help to shed new insights on many key issues discussed in
the preceding text and that will be dealt with in greater detail in the
contributed chapters of this volume.

1.3.1 Economic patterns

Liangzhu society was heavily reliant on rice farming. The enormous
volume of rice discovered in the storage pit at the Mojiaoshan platform
(see Chapter 2 for more details) provides unambiguous evidence for the
large-scale consumption in the Liangzhu elite circle. However, there is no
evidence for rice production inside the walled area of the Liangzhu centre.
Rather, most evidence of Liangzhu-period rice farming comes from outside
the centre, including the excavation at the well-preserved paddy field site of
Maoshan, the more commonly encountered farming and harvesting tools
during excavations, and, more directly, archaeobotanical remains discovered
through flotation.

Before excavation, a survey conducted by Zheng Yunfei and colleagues
compare phytolith concentrations in soil samples collected from an area
of 14 ha at and around the Maoshan site. Of the 80 analysed samples, 52
samples contained abundant rice phytoliths (minimum 5,000 phytoliths
per/g, maximum 260,000 phytoliths per/g). Based on these results, they
reconstruct the distributional scope of the paddy field to around 5.5 ha
(Zheng et al. 2014). The enormous scale of the field is confirmed through
the excavation of a small part of the field. The paddy fields were built
and used in two phases, and the transition from the early to the late
phases was accompanied by a clear expansion of the single paddy unit
size (see Chapter 2 for more information) and intensifying farming
practices (Zhuang et al. 2014). The location of the Maoshan paddy field is
ideal, located in the intermediate area between the foothills and the alluvial plain, and advantageous in terms of drainage. These environmental
conditions are found at many other small to medium settlements in the
region and it can be surmised that these settlements were the basic production units for rice farming in Liangzhu society. Abundant carbonised
rice grains were discovered in the burned-down deposit of storage (c.
0.07 ha) near the Mojiaoshan palatial platform. This deposit contained
10,000 to 15,000 kg of rice and potentially more when the pit was fully
filled with rice (see Chapter 2). Considering the lack of evidence for rice
farming at the Liangzhu centre, this ruin of storage facility provides evidence that the centre probably received rice supplies from the production
units just mentioned.

Cross-referencing other economic sectors apart from the jade industry,
such as the ceramic industry and stone tool production, the economic

relationship between the Liangzhu centre and its neighbours becomes clearer. Chapter 5 briefly discusses the ceramic industry of Liangzhu. Several distinctive characteristics are worth reiterating here. (1) The production of black pottery with carved patterns appear very similar to those appearing on jade, as suggested by Zhao and others (e.g., Lu et al. 2013), representing a pronounced technological leap forward from the preceding Songze Culture period, which made possible the transformative technologies brought out by advanced jade crafting and complicated design. (2) The similarity of the black polished pottery found at the Liangzhu period sites across the Taihu Lake region (and beyond) in terms of the design and production technologies indicates that there was possibly a production centre(s) for this kind of special ceramics or the local potters were receiving or influenced by ideas or information from the centre, either directly or indirectly.

Was there economic specialisation in Liangzhu society? We know that the production of high-quality jade was definitely monopolised by the elites at the Liangzhu City and other subregional centres who had access or direct control of this rare raw material and mastered the crafting skills. Some of the chapters in this volume are dedicated to further elucidating this pattern. Hence, current evidence shows that these elites did not cultivate rice by themselves. To perform their time- and energy-consuming jade crafting, they had to rely on external suppliers for their food. Thus, we can see the existence of an asymmetric economic relationship between the Liangzhu City and its neighbours. The difficult questions facing archaeologists is how this asymmetric economic relationship related to the economic specialisation in the wider Liangzhu region and how this helped to shape the political structure at the centre.

The Uruk-Warka in Mesopotamia has long been considered a centre with a 'centralized and specialized production of mundane, utilitarian goods' with 'an administered system of exchange' (Pollock 2001, p. 183). Algaze further suggests that the craft production (e.g., 'standardized ceramics' and metal working) at Uruk-Warka was 'on an industrial scale' (ibid., p. 35), which is also supported by discovered historical texts and scripts (Nissen et al. 1993). Alternatively, however, some scholars advocate a much less centralised economic structure in Uruk Mesopotamia and that the economic specialisation is an outcome of differentiated availability in resources among different regions. Utilitarian goods were produced within local communities and it was the economic force of local production that played a pivotal role in creating the asymmetric economic relationship between Uruk-Warka and its neighbours (Adams 1981). Data from recent surface surveys has supported this model. Although focusing on Ubaid period ceramics, Berman's scientific study of ceramics is relevant to the study of Uruk-period pottery. She demonstrates that ceramic products at local sites come from various production centres, 'none of them under centralised control' (in Matthews 2003, p. 107). This diversity in production is at odds with, according to Matthews, 'the great stylistic homogeneity of pottery across the Ubaid world' (ibid.).

The critical issue is whether the ceramic industry is an integral part of an 'integrated economy', to use a term by Pollock (2001, pp. 208–209), who argues strongly, based on her full-coverage survey data, that at least in some parts of southern Iraq that they did not have a 'highly integrated economy'. The seemingly homogeneous style of the black pottery with carved symbols, mentioned previously and examined by Zhao in this volume, might point to a central organisation of production and/or distribution. However, this should not be confused with the functions and cultural meanings of pottery. The collective identity or ideology within Liangzhu society was consolidated by the central enforcement and the limited use of distinctive Liangzhu jade. This is different from the site of Uruk in Mesopotamia that did not deploy jade or similar objects to mediate a centralised identity. It might then be tentatively suggested that the production of certain types of pottery at Liangzhu was centrally organised as part of the integrated economy focused on jade production. Contrary to this was the more independent production and organisation of the pottery industry versus its more important socio-economic role in shaping the ideological identity in Uruk, Mesopotamia (Berman 1994, p. 29, cited in Matthews 2003). This Liangzhu model should be tested in future research such as using full-coverage surveys that provide more spatial information regarding where the pottery appeared in the site complex and its density.

1.3.2 Labour organisations

Liu et al.'s contribution briefly discusses the scale of labour organisation and logistics for the construction of the earthen works at the Liangzhu City. Elsewhere, we have speculated about the productivity of the construction works. In this section, we further discuss the organisational structure of the labour needed and its wider socio-economic implications within a comparative perspective. Poverty Point is an enormous Archaic period site (c. 1500 BC) in the United States in northeast Louisiana. The 3 km² area consists of six mounds and several concentric earthen ridges (believed to be used for astronomic observations) (Kidder 2002). Contrary to conventional ideas that advocate a universal link between early urban centres and stratified agrarian societies, the Poverty Point compound is considered a 'great town' or a 'cosmopolitan' society (Sassaman 2005, p. 336) constructed and occupied by a group or groups of people who were hunter-gatherers with 'hereditary leaders, sedentary lifestyles, and complex social, political and economic relationships' (Ortmann and Kidder 2013, p. 66). Drawing together different evidence, Sassaman suggests that the construction of the site was 'perhaps local and corporate, the work of a collective body'. But to achieve this,

Sassaman goes on indicating that, based on the existence of abundant non-local materials, the Poverty Point residents were able to mobilise not only external resources but also perhaps 'other people' (Sassaman 2005, p. 336). Some recent research provides more insights into the organisation of the construction of individual mounds at the site. The short chronological span of the main mounds has led many scholars to believe that the construction of the earthen works was a rapid process. Mound A, a massive mound (207 m × 210 m) with steep sides, was estimated to have been built in 90 days by 'a large group of hunter-gatherer laborers' who 'were simultaneously engaged in the project' (Ortmann and Kidder 2013, p. 78). Large amounts of resources were pulled together to move the 765,000 m³ of earth and build the mound. Comparing different energetic estimates on labour productivity and duration of the construction of Mound A, Ortmann and Kidder suggest that 91,700 person-days (five-hour days) of labour would be needed to dig and pile the c. 238,500 m³ of earth required to build the mound, excluding the calculations for other costs such as transportation. They do not support, however, the idea that 'the site supported a permanent population of up to 9000 individuals'; rather, the 'construction was the product of a modest to moderately large, local, permanent resident population' (ibid., p. 76).

The implications for this research on the construction process of Poverty Point to our comparative discussion here are twofold. First, how was the labour mobilised and how were the logistics organised by this group of hunter-gatherers? Was the construction seasonal or year around, and how was it related to the temporality of food production at and near the site? How much effort was spent in mobilising the labourers and through what mechanism was this achieved? Gibson (2001) has suggested that the labourers for the construction of the mounds at Poverty Point were free labourers who were drawn to the construction through a sharing and reciprocal relationship between the Poverty Point community and its neighbours. There is clear material evidence (e.g., the common occurrence of non-local products at Poverty Point) for the material flux within the region. A full explanation will not be possible until finer resolution archaeobotanical analyses regarding the seasonality of food production and storage have been undertaken. Second, even though the construction of several mounds at Poverty Point were completed in rapid succession and probably under a central organisation and planning, the constructional hiatuses identified through geoarchaeological surveys and 14C dates indicate that the construction of the whole site was piecemeal and represents multiple events over different time periods.

The first and foremost point to make in returning to the Liangzhu case is that the Liangzhu is a rice-based economy. This means that the seasonal availability of labourers (presumably rice farmers) would have been different from that at Poverty Point. There are certain seasons of the year when rice farmers are occupied in the field, such as when they are sowing rice in the spring and harvesting rice in the summer or autumn. Under the assumption that rice farmers were the main source of labour for the construction of the Liangzhu City and related infrastructures outside the city, rice farmers likely only worked as seasonal labourers. If this was the case then the rice farmers, living outside of the Liangzhu City, were mobilised during their 'free time' away from the rice paddies to work at the centre; we may ask, how was this labour force mobilised and organised? Were these rice famers all from the groups or communities living outside the palatial compound in the Liangzhu City, immediately outside the city, or from further away? Was there a rotating schedule between different groups? All these questions remain to be answered. Future research might be able to tease out detailed information regarding the seasonality of the construction projects preserved in, for instance, the earth and grasses used in the earthen works.

1.3.3 Interactive spheres

A brief comparison with established cases of centre-periphery interactions at other sites in the ancient world could shed new insight into the so-called urban/rural division at and around the Liangzhu City (cf. Zhao 2017). The establishment of the exploitative economic relationship between Uruk and its neighbours benefited from and, to a greater extent, was determined by Uruk's geographic advantage in Southern Mesopotamia for transportation. Uruk-Warka, which lacked many key resources for economic production (e.g., metals and timber), could exploit its northern neighbours through its convenient location for transportation (Algaze 2001). This south-north interaction played a key role in the rapid growth of Uruk during the fourth millennium BC. Uruk's success lies fundamentally in its political and economic institutions centred around the 'institutionalized rulers whose power was based on religious, political, and military roles' (ibid., p. 34). This powerful statement by Algaze has been challenged by other scholars who have demonstrated that wider and more complicated local networks were in place and contributed to the economic developments in Mesopotamia. Differences in this exploitative economic relationship between the Uruk and its neighbours seemed to have formed a reciprocal relationship with its neighbours, with Uruk appearing to have coercive control over its neighbours' resources. Poverty Point, in terms of the scale of its monumental structure, was unique or 'not replicated elsewhere', as Sassaman puts it (2005), in the region. This established new social order in the region had a primary focus on 'materialising' symbolic power (ibid.).

Before we examine the Liangzhu case, let us look at some other case studies in north China and the Middle Yangtze River, contemporary with or slightly later than Liangzhu. As briefly discussed in section 1.2, one of the radical social changes that took place in late Neolithic China was the rise of walled sites or towns in different regions. Some of the aforementioned sites such as Taosi and Shijiahe emerged as regional centres, but as some scholars have pointed out (cf. Zhao 2017), these sites were often surrounded by several sub-centres with relatively large size and complex settlement structures. A very good example is the Shijiahe Site Cluster and its contemporaries (Zhang, C. 2013). The Liangzhu Site Cluster might be different from Shijiahe in terms of its position in the regional settlement patterns. Similar to the case at Poverty Point, one of the most salient characteristics of the Liangzhu City is the strong link established with its neighbours through jade and the homogeneous decorative system. There has been much debate in terms of the importance and use of military power by the Liangzhu elites and how they used it to 'materialise' their control over the region (see Chapter 5 and section 5.2). If we agree with this model, it would indicate that the symbolic relationship between the Liangzhu City and its neighbours was realised through the former's control over the latter, underpinned by the economic exploitation of the centre on the peripheries. This is mirrored by what we have seen in the patterns of rice production and consumption at Liangzhu.

To conclude, the comparison of Liangzhu with other prehistoric urban centres provides many useful insights as well as raising many more questions. A coherent, holistic framework should be developed to better integrate the economic, technological, and ideological aspects of the Liangzhu City and its relationship with its neighbours. We hope that the first-hand data and primary models presented and articulated in this book will facilitate more researchers embarking on this exciting academic journey.

2 The Liangzhu City

New discoveries and research

*Liu Bin, Wang Ningyuan, and Chen Minghui,
translation by Catherine Xinxin Yu*

2.1 Liangzhu Culture and world civilisations

Discovered in 1936, the Liangzhu Culture was at first seen as a branch of the Longshan Culture from Shandong Province that disseminated towards the Lower Yangtze River Region. As more archaeological evidence became available, the distinctive characteristics of this culture were gradually recognised and formally named in 1959. With the discovery of large-scale elite cemeteries including Fanshan and Yaoshan in Zhejiang Province, Fuquanshan in Shanghai, and Sidun in Jiangsu Province in the 1980s, the importance of the Liangzhu Culture among prehistoric Chinese archaeological cultures became clearer. The difference in rank between large and small burials and the abundance of ritual jade such as *cong* tubes, *bi* disks, and *yue* axes elucidated the role of religion and power in the Liangzhu Culture, revealing a highly developed social organisation already in place in Liangzhu society.

Earthen structures with stone foundations that resemble city walls or river dikes were discovered in 2006 when the Zhejiang Provincial Institute of Cultural Relics and Archaeology excavated the Putaofan site, east of Pingyao Town. After conducting systematic surface surveys and excavations for almost a year, we discovered and defined the boundaries of the Liangzhu City that extends 1,900 m north to south and 1,700 m east to west, occupying an area of approximately 300 ha. This opened a new chapter in the study of the Liangzhu Culture. The discovery of the Liangzhu City offered reliable evidence for the distribution of more than 300 sites and locations around Liangzhu Town and Pingyao Town and sheds new light on how advanced the Liangzhu Culture was at this point. It has since become a commonly accepted understanding within the archaeology community within China that the Liangzhu Culture was an advanced civilisation.

Many natural and cultural wonders are located in the so-called magical zone around latitude 30°N, including the ancient Egyptian civilisation along the Nile River, the Sumerians in Mesopotamia, and the Harappa civilisation in the Indus River valley. The core area of the Liangzhu Culture around the Taihu Lake region in the Lower Yangtze region is located at 30°–32°N and 119°10″–121°5″E, similar to the world-renowned cradles of civilisations listed previously. The Taihu Lake region occupies an area of 36,500 km², bordering on the Maoshan Mountain and Tianmu Mountain to the west, the Yangtze River to the north, the Qiantang River to the south, and the East China Sea to the east. Surrounded by mountains, criss-crossed by rivers and dotted with lakes, this fertile area was highly suitable for human habitation.

The area around the Taihu Lake region is culturally distinctive and has a clear sequence of Neolithic cultures. The sequence of the Majiabang Culture, the Songze Culture, the Liangzhu Culture, the Qianshanyang Culture, and the Guangfulin Culture from 5000 BC to 2000 BC shared the same cultural roots and maintained cultural continuity while evolving and eventually developing into the Wu-Yue Culture of the pre-Qin era (before 221 BC). The period when the Liangzhu Culture flourished (3300–2300 BC) coincided with active regional interactions in prehistoric China, when the Miaodigou Culture from the Yellow River valley and the Songze Culture from the Yangtze River Delta expanded and integrated, and archaeological cultures from every region gradually moved towards a stage of steady but rapid development. Social differentiation could be clearly manifested in the material culture. As a typical example, the Liangzhu Culture created a sacred image that was unanimously acknowledged by this society for its symbolic significance and designed a system of ritual jade with items such as *cong* tubes, *bi* disks, *yue* axes, and cockscomb-shaped *guanzhuangshi* objects for divine worship. Exclusive control of kingship, military power and finance was achieved through the control of divine power. Elaborate Liangzhu burials with a large amount of ritual jade demonstrate the prestige of rulers and gender division of labour among the elite classes. The ritual jade system and the idea of the divine right of rulers created by the Liangzhu Culture was inherited and further developed by subsequent Chinese cultures. The period when the Liangzhu Culture actively coincided with the emergence of Ancient Egypt, Sumer, and Harappa and was an important period during which both Chinese and world civilisations emerged and developed independently. As such, we believe that the period represented by the Liangzhu Culture can be called the Liangzhu period. It was a continuation of the Miaodigou period (Han 2012; Chen 2013) and Songze period, and succeeded by the Longshan period.

2.2 Location of the Liangzhu site cluster

Referred to as the Liangzhu 'capital' by Yan Wenming (Yan 1996), the Liangzhu site cluster was at the heart of the Liangzhu Culture. The Liangzhu site cluster and Liangzhu jade objects are the most representative material evidence from the Liangzhu Culture.

The Yuhang District of Hangzhou City in Zhejiang Province is situated between the mountainous zone of western Zhejiang and the plains of northern Zhejiang. It is surrounded by the Tianmu Mountain to the west and by its subranges to the north and south but is open to the east, forming a

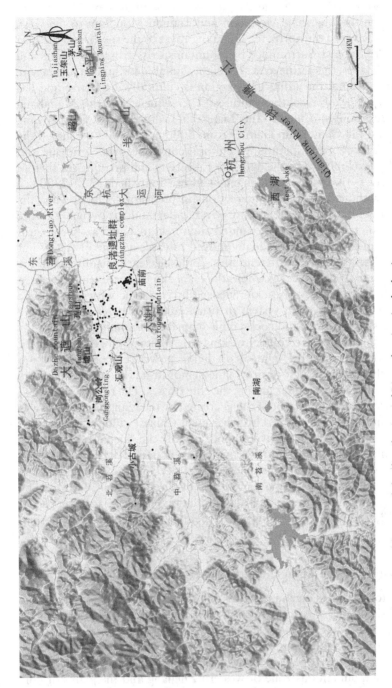

Figure 2.1 The C-shaped basin and the distribution of Liangzhu sites in the basin

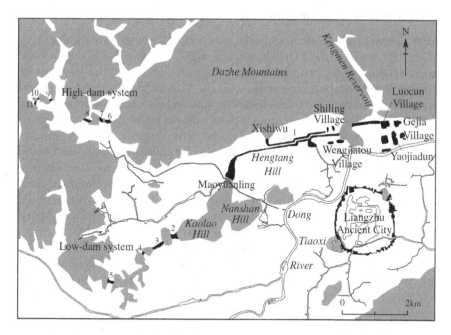

Figure 2.2 Location of the Liangzhu site cluster and important excavation locations
mentioned in the chapter, including the hydraulic system outside the
city. Low dams: 1. Tangshan; 2. Shizishan; 3. Liyushan; 4. Guanshan;
5. Wutongnong; high dams: 6. Ganggongling; 7. Laohuling; 8. Zhoujiafan;
9. Qiuwu; 10. Shiwu; 11. Mifenglong

C-shaped basin that stretches for 42 km from east to west, and 20 km from
north to south, occupying a total area of 1,000 km². The Banshan Mountain,
the Chaoshan Mountain, and the Linping Mountain to the north-east of
Hangzhou are located near the opening of this C-shaped basin, separating
this plain from the rest of the landscape (Figure 2.1).

The Liangzhu site cluster is located in the northern part of the C-shaped
basin, about 20 km north-west of the city centre of Hangzhou. The Dazheshan
Mountain and the Daxiongshan Mountain, two mountains in the subrange
of the Tianmu Mountain range, tower over the southern and northern sides
of the Liangzhu site cluster, while the Huiguanshan Mountain, the Yaoshan
Mountain, the Nanshan Mountain, and Kaolaoshan Mountain are scattered
to the west, approximately 2 km from the site. Towards the east is an open
plain. This choice of location gives the impression that the mountains were
used as walls. The Dongtiao River, originating in the Tianmu Mountain
range, meanders to the north-east and turns north once it reaches Zhangshan
Mountain, eventually feeding into the Taihu Lake at the end. In addition to
nourishing the land, it also acted as the most convenient waterway from the
Liangzhu site cluster to the Taihu Lake region and beyond (Figure 2.2).

2.3 Layout of the Liangzhu City

After eight years of excavations, surface surveys, and coring since the discovery of the city walls of the Liangzhu City in 2007, we have established a basic understanding of its structure and development. The Liangzhu City can be divided into three concentric zones with the Mojiaoshan Palatial Compound in the centre, surrounded by the inner city wall and the outer circle. Distinction in status between these three zones is clearly shown by the decreasing height of the artificially built earthen platforms as one moves from the centre to the outer limits of the city. This tri-circle urban structure is similar to capital cities of later periods that have a palace district, a royal city, and an outer city. The Liangzhu City is thus significant for being the earliest example of this type of three-circle city in China. To the north and north-west of the Liangzhu City, there is also a large-scale hydraulic system as well as the Yaoshan and Huiguanshan 'Altars' that were probably used for astronomical observation.

We have established the chronological framework of various sites inside the Liangzhu City. The Fanshan elite cemetery north-west of Mojiaoshan inside the city and the elite cemeteries on the Yaoshan and Huiguanshan altars outside the city to the north-east and north-west date to around 3000 BC. Organic remains discovered in six of the ten newly discovered dams north-west of the city, including Ganggongling, Liyushan, Shizishan, Laohuling, Zhoujiafan, and Qiuwu, have been Accelerator Mass Spectrometry (AMS) dated by the Chronology Laboratory of the School of Archaeology and Museology at Peking University to the Early Liangzhu period, around 3100–2850 BC.

Abundant rubbish deposits from the late Liangzhu period were discovered in the moat outside the city wall and on the slopes of the Mojiaoshan platform, demonstrating that these locations were important residential areas during the late Liangzhu period. No direct evidence has yet been found regarding the *terminus post quem* of the construction of the city wall and the Mojiaoshan platform, but based on the consistency of the layout and design of the Liangzhu City, we believe that they should date to around 3000 BC, close to the construction date of the Fanshan and Yaoshan cemeteries and the hydraulic system (c. 3100–2900 BC) outside the city.

Archaeological excavations conducted at Wenjiashan, Bianjiashan, Meirendi, Lishan, and Biandanshan within the outer circle show that their construction was gradually undertaken and completed between the middle and late Liangzhu period.

2.3.1 City walls and layout of the city

The Liangzhu City has a rectangular shape with rounded corners and situated along a north-south axis. It is 1910 m long north to south, 1770 m wide east to west, covering a total area of nearly 300 ha. The city wall is approximately 6 km long, 20 m to 150 m wide, and 4 m high at the best preserved sections. It incorporates the naturally formed Fengshan Mountain

Figure 2.3 Section of northern (left) and southern (right) Liangzhu City walls

and Zhishan Mountain at the south-west corner and north-east corner, respectively. A layer of rock foundation, about 0.2–0.4 m thick, is laid at the bottom to reinforce the rammed earth wall made with yellow earth collected from the nearby mountains. This method of building yellow earth walls with stone foundation is unprecedented in contemporary sites in China and abroad (Figure 2.3).

All sides of the city wall are sandwiched between an inner and an outer moat, except the southern side where there is no outer moat. Eight water gates connecting the inner and outer moats have been found through coring, two on each of the four sides of the city wall. The water gates are usually 30 m to 60 m wide, but the two on the western wall are narrower because they face the Tiao River, one 10 m wide and the other 20 m wide. The ground along the southern wall is higher and no outer moat has been found, but an overland gate made with three little earthen platforms is located in the central section of the southern wall. The three earthen platforms are not interconnected but positioned symmetrically, forming four gateways that lead in and out of the city.

While the city wall was being constructed, the city centre was also planned and constructed as a whole. Many artificial earthen mounds were distributed throughout the city, which were mainly used as foundations for buildings and as cemeteries. Mojiaoshan, the palace district inside the city, is situated on such an earthen mound in the centre of the city, occupying a tenth of the city's total area. The Fanshan elite cemetery, where 11 high-status burials with an abundance of Liangzhu jade objects were excavated in 1986, is situated right next to the north-west corner of the site of Mojiaoshan. In addition to the moats along the city wall, there are also 51 canals and rivers inside the city. There are main canals located north, east, and south of Mojiaoshan, following a layout with a shape resembling the Chinese character 工. Tributary canals were dug out, linking the main canals to the inner moat, to form a grid pattern. This criss-crossing network of canals formed a complete water transport system, making the Liangzhu City a city on water because canals were the most important method of transport. The survey

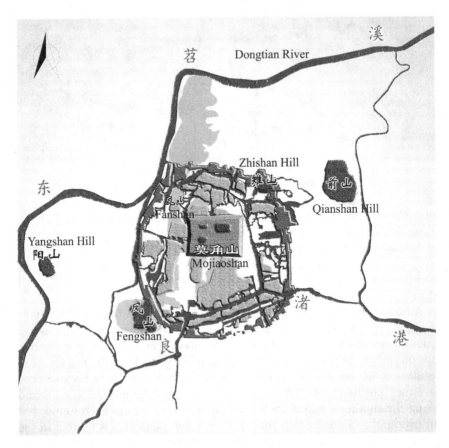

Figure 2.4 Landscape around the Liangzhu site cluster: early phase

results indicate that many of these canals and moats were dug up purpose-fully, and the excavated earth was used to construct earthen mounds such as the one on which Mojiaoshan is situated. With 9,538 m of moats, 11,733 m of canals inside the city, and 10,291 m of canals in the outer circle, the total length of the city's canals adds up to 31,562 m, in other words, exceeding 30 km. The width of the canals varies between 10 m and 50 m, with a depth ranging from 2 m to 4 m.

We roughly divided the earthen platforms inside the city to an early and a late phase based on the stratigraphic evidence which revealed that platforms and canals overlap. Early-phase earthen platforms are mainly distributed along the northern, western, and southern sides of Mojiaoshan and along the banks of the main canals and the inner moat. Most of them took advantage of the natural features of the land with high topography. Based on our survey data, we estimate that there are 37 such early earthen platforms (Figure 2.4).

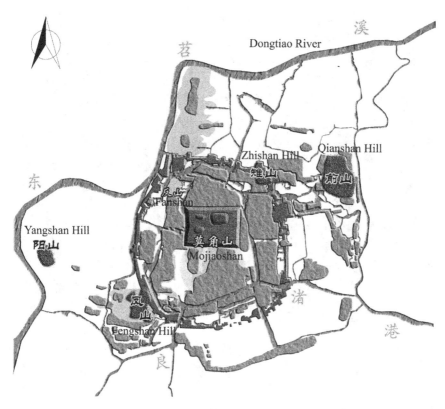

Figure 2.5 Landscape around the Liangzhu site cluster: late phase

There were more canals during the earlier period but many were filled in by rubbish deposits over time. Some were even covered by a layer of yellow earth and converted into residential areas. Therefore, there were considerably fewer small canals during the late Liangzhu period, while residential areas increased drastically. We estimate that there are 19 earthen platforms from the later period. In addition to the expansion of residential areas, the city walls were also densely inhabited due to population increase during the later period. At the same time, residential areas were also built outside the city, to the north, east, south, and south-west, forming the outer circle (Figure 2.5).

2.3.2 Mojiaoshan platform

Mojiaoshan platform, the earliest urban palatial district in China discovered so far, is located in the centre of the Liangzhu City. It is an artificially constructed earthen platform with a rectangular shape. The base of the platform is 630 m by 450 m, occupying an area of nearly 30 ha. Our estimation

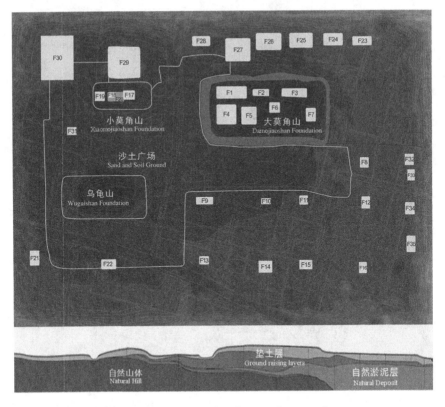

Figure 2.6 Plan and diagram of the earthen layers of Mojiaoshan

suggests that 2,280,000 m³ of earth would have been moved to build the platform. Coring surveys show that the western part of Mojiaoshan platform was built on a small, natural hill. Greyish mud from the nearby marsh was dug out and used to fill and raise the lower ground east of the mount in order to form the overall base of the platform; then, yellowish earth was piled on top. This artificial mound is 10 m to 12 m deep on the east side and 2 m to 6 m deep on the west side (Figure 2.6).

There are three smaller earthen platforms – Damojiaoshan, Xiaomojiaoshan, and Wuguishan – on top of the Mojiaoshan platform. They are likely the foundations of large-size palaces. Damojiaoshan, the largest and tallest of the three small platforms, is situated on the northeastern side of the Mojiaoshan platform. The platform has a rectangular dome shape and measures 180 m by 97 m at the base. It is 18 m above sea level, with an artificial earthen buildup of 16.5 m and a relative height of 5 to 6 m. During the excavation seasons from 2013 to 2015, we discovered seven foundations for raised-platform architecture arranged in two rows on top of Damojiaoshan, each with an area of about 0.03–0.09 ha (Figure 2.7).

Figure 2.7 Earthen-platform architecture no. 1 on top of Damojiaoshan

Excavation shows that ditches, about 5.5 m to 12.8 m wide and 0.6 m to 1.5 m deep, surrounded Damojiaoshan. They were left sunken at the initial stage when the land was filled with greyish mud but were later filled after a period of use. Criss-crossing wooden planks pieces were placed between the yellowish brown earth and the greyish earth on the northern and southern slopes of Damojiaoshan. It is likely that these wooden planks were placed on top of the greyish layer of earth at the base of the Damojiaoshan platform during construction to evenly distribute the weight and reinforce the entire structure (Figures 2.8 and 2.9).

Xiaomojiaoshan is located west of Damojiaoshan. It measures 90 m by 40 m at the base, 17 m above sea level, with a relative height of 5 m and an artificial earthen buildup of 6 m. Wuguishan is located south of Xiaomojiaoshan. The top of Wuguishan as well as the western and northern slopes have been damaged by recent earth removal by local villagers. Structures on top of the platform have also been completely destroyed. The platform measures 130 m by 67 m at the base, 16.5 m above sea level, with a relative height of 4 m and an artificial earthen buildup of 7 m.

In addition to the three platforms, other large structures at Mojiaoshan include extensive areas of rammed earth and sand, a burnt red clay deposit, remains of houses, and stone wall foundations. Rammed earth and sand areas are widely distributed between Damojiaoshan, Xiaomojiaoshan, and Wuguishan as well as south of Dmojiaoshan and Wuguishan, occupying a total area over 7 ha. They were created by pounding alternating layers of sand and clay and are found on top of the yellowish brown layer of earth. The thickness of the rammed earth

Figure 2.8 The ditch on the south slope of Damojiaoshan and wooden planks at the bottom of the ditch

Figure 2.9 Remains of stone structures at Damojiaoshan

and sand layer is generally 0.3–0.6 m, but the thickest part reaches 1.3 m, where obvious pounding marks can be seen (Zhejiang 2001). The area with rammed earth and sand should have been an important area for major activities on top of the Mojiaoshan platform. Red burnt earth deposits occupying nearly 0.5 ha on the southeast slope of the Mojiaoshan platform are probably the remains of burnt houses. Stone foundations, found all around Damojiaoshan as well as in the east and north corners, were probably covered ditches initially used for drainage. The remains of three large ditches have also been found south of Damojiaoshan. They are mainly north-south–oriented ditches with a small number of east-west–oriented ditches. Pieces of wood oriented north to south were probably placed inside the ditches originally. Based on the discoveries on the south slope of Damojiaoshan and Meirendi, it is possible that there were wooden planks oriented east to west placed on top of wooden planks oriented north to -south, creating a surface for outdoor activities (Figure 2.10).

A wooden structure was found in the construction deposit on the southwest slope of the Mojiaoshan platform. This area was initially a canal. When the canal was no longer in use, it was filled with clay wrapped up by grasses to form a new bank. Wooden poles were locked tightly in the clay layer and acted as reinforcement (Figure 2.11). Excavation of the east slope of Mojiaoshan shows that the yellowish clay layer was built gradually with patches and ridges of clay, identical to the construction technology applied in the south city wall, which might also have been built by piling clay wrapped in grasses. Also, the pit no. H11, where 13,000 kg of carbonised rice was deposited, has been cleared (Figure 2.12).

2.3.3 The outer circle

There are long strips of artificially constructed highlands in the outer circle of the Liangzhu site cluster, including Biandanshan-Heshangdi, Lishan-Zhengcun-Gaocun, Bianjiashan, Dongyangjiacun, and Xiyangjiacun, each about 30 m to 60 m wide, with artificial earthen work buildup of about 1 m to 3 m. These intermittent strips create various frames around the ancient city wall and form the basic structure of the outer circle. They stretch 2,700 m from Biandanshan to Bianjiashan, 3,000 m from Lishan-Zhengcun to Zhangjiadun, occupying a total area of 8 km^2. Long strips of residential areas such as Meirendi, Zhongjiacun, and Zhoucun are distributed in the area between the highlands and the city wall. This kind of residential strips are rather densely distributed to the north-east and south-west of the city. Dongyangjiacun, Xiyangjiacun, Dushan, Wenjiashan, and Zhongjiashan form two to three incomplete rings surrounding the south-west side of the Fengshan Mountain (Figure 2.13; see also Figure 2.6). In the north-west, Zhoucun, Meirendi, Qianshan, Zhishan Mountain, and Majinkou also form a small ring that surrounds the north water gate of the east city wall and the river that runs through this gate. These two 'framing' structures are important components of the outer circle and demonstrate that the inhabitants of the Liangzhu City carefully considered the advantages of nearby hills such as Fenshan, Zhishan, and Qianshan

Figure 2.10 Remains of ditches south-west of Damojiaoshan

Figure 2.11 Remains of the wooden structure and clay wrapped by grasses on the south-west slope of Mojiaoshan. 2.11-1: Wooden structure; 2.11-2: Wooden structure and clay wrapped by grasses

while designing residential areas. The presence of the outer circle shows that designated zones outside the city wall were planned and used as residential areas. They were integral parts of the Liangzhu City.

Archaeological excavations of various scales have been undertaken at sites located in the outer circle, including Wenjiashan, Zhongjiashan, Dushan,

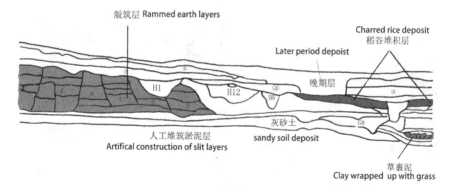

Figure 2.12 Illustration of the stratigraphy of the east slope of the Mojiaoshan Platform

Figure 2.13 Digital elevation model (DEM) of the periphery of the Liangzhu site cluster

Figure 2.14 Remains of the wooden riverbank at the Meirendi site. Excavation photos taken from different angles

Figure 2.15 The dock at the site of Bianjiashan

Bianjiashan, Meirendi, Lishan, and Biandanshan. Excavations showed that Meirendi, Lishan, and Biandanshan are raised residential areas that were built up during the late Liangzhu period. Remains of the wooden riverbank on the edge of Meirendi earthen platform sketch a picture of waterside living around 4,500 years ago (Figure 2.14) (Zhejiang 2015a). However, cemeteries, ancient canals, and docks from the middle to late Liangzhu period are discovered at the Bianjiashan site (Figure 2.15) (Zhejiang 2014). A middle to late Liangzhu period cemetery was also found at Wenjiashan (Zhejiang 2011).

2.3.4 Hydraulic system

A large hydraulic system north-west of the Liangzhu City has been discovered through coring surveys and small-scale excavations in recent years. The 11 dams discovered so far are located in front of the valleys formed by two mountains and can be divided into north and south groups. Lower dams including Tangshan, Shizishan, Liyushan, Guanshan, and Wutongnong make up the south group, while higher dams including Ganggongling, Laohuling, Zhoujiafan, Qiuwu, Shiwu, and Mifenglong make up the north group, creating a protective system with two lines of barriers (Figure 2.2) (Zhejiang 2015b). The reservoirs of this hydraulic system located north and north-west of the Liangzhu site cluster occupies an area of 1,240 ha and has a capacity of 60,000,000 m³, based on our computational simulations.

The Tangshan site is spread out along a hill north of the Liangzhu City. It is 5 km long, 20–50 m wide, with a relative height of 2 m to 7 m. It is located about 100 m to 200 m from the foothills of the Dazheshan Mountain. The top of the dam is 12 m to 20 m above sea level. The middle part of the Tangshan levee has a double-section structure where an east-west–oriented canal is found between the top and bottom sections of the levee. The Shizishan, Liyushan, Guanshan, and Wutongnong dams, west of the Liangzhu City, are about 35 m to 360 m long, 100 m wide, 10 m above sea level, with an artificial buildup of 10 m. They are aligned south-west of the Tangshan levee.

Ganggongling and the other high dams are 8 km north-west of the Liangzhu City. They are about 50–200 m long, 10–20 m high, and 25–40 m above sea level. Ganggongling was built using a similar method as the construction of Mojiaoshan, with greyish clay being raised at the bottom, overlain by yellowish silty clay on the top (Figure 2.16).

The hydraulic system is made up of a series of dams connected by natural hills and mountains. The Tangshan levee was built with a stone base and clay piled on top, just like the city wall of the Liangzhu City. By contrast, high dams like Ganggongling and Mifenglong have a mud core covered with yellowish clay on the outside, just like the Mojiaoshan platform. The core of dams like Ganggongling, Qiuwu, Wutongnong, and Shizishan is built with clay wrapped up with grasses. Therefore, the structure and building techniques of these dams are identical to typical sites in the Liangzhu site cluster, which represents significant evidence that these dams belong to the Liangzhu Culture.

The structure and function of this hydraulic system is not yet fully understood as systematic investigation has only just begun. Based on field observations, we think that the function of this hydraulic system includes flood prevention, transport, water storage, and irrigation (Wang and Yan 2014).

The Tianmu Mountain is the location of the heaviest rainfall in Zhejiang Province where summer mountain torrents are prone to cause direct damage

Figure 2.16 Section of Ganggongling dam damaged by soil removal

to the downstream plain where the Liangzhu site cluster is located. The hydraulic system can relieve the threat of flooding by diverting a large amount of incoming water within the natural valley and the depression between high and low dams. Studies have shown that these dams can resist about 870 mm of continuous precipitation fallen within a short period of time, an amount comparable to the maximum local precipitation that occurs as rarely as once in a century.

This hydraulic system is also very useful for transport. The Tianmu Mountain range can provide rich sources for stone, timber, and other animal and plant resources to nearby settlements, but unlike plains with developed water transport networks, the natural conditions are not favourable for water transport in this area. Mountains and valleys are steep and seasonal variation in rainfall means that there might be mountain torrents in summer but that rivers may dry up in winter. The water stored in the dams can connect water transport routes in each valley to form a connected transport network. Take tall dams such as Ganggongling, Laohuling, and Zhoujiafan as examples. Our calculation, using the lowest dam height of 25 m above sea level, shows that they can increase the water level further up in the valley by about 1,500 m when the water is at full capacity. Low dams, such as Liyushan dam, have a height of about 9 m above sea level and can extend northward for about 3,700 m when the water is at full capacity, reaching the base of Ganggongling. The storage volume of the high and low reservoirs would have been about 1,498 and 5,072 m³, respectively (Liu et al. 2017). This colossal figure, albeit based on crude calculations, convincingly demonstrates the scale of landscape transformation by Liangzhu society. Further research should aim to unpack how this hydraulic landscape functioned for irrigation, water transportation, and other purposes, all being prominent and important features of early states (Algaze 2005, 2009).

In historical documents, the history of water management in China usually starts from 2000 BC with the legend of Gonggong, Gun, and Da Yu (the Great Yu) managing the great flood (widely known as the Greater Yu's Taming the Floods), but there are no known archaeological remains of water management systems from that period. The archaeological remains of the earliest large-scale water management systems discovered so far are dated to the much later Spring and Autumn or Warring States periods. However, hydraulic facilities such as Tangshan and Ganggongling near the Liangzhu City can be dated to c. 3100–2900 BC, making them the earliest example of large-scale water management systems in China. The close connection between this hydraulic system and the earliest urban centre in China, the Liangzhu City, provides a significant example of ingenious landscape engineering and construction for effective water management and transportation. This is very important in the history of urban planning and construction in China and worldwide.

Among other contemporary ancient civilisations, ancient Egypt also built dams quite early on as well. The main function of their dams was to minimise the damage of the periodic flooding of the Nile. Although some dams also facilitated irrigation and transport, very few remains of this type of dam have been found so far. During the Old Kingdom, in 2650 BC, ancient Egyptians built Sadd el Kafara dam on Wadi Garawa, 30 km south-east of Cairo. Its function was to barricade the winter mountain torrents from the east and collect water to form a permanent reservoir. The two sides of the dam were about 24 m wide and were constructed with limestone ashlars, 0.3 × 0.45 × 0.8 m in size. The upstream side is sloped at about a 30° angle. The core of the dam was about 36 m thick, filled with clay and pebbles. The dam was 113 m long and 14 m high, with a capacity of 500,000 m³. It took 10 to 12 years to construct (Garbrecht 1985). Its size is roughly comparable to the Ganggongling dam of the Liangzhu hydraulic system. Though irrigation canals were built very early in Mesopotamia (Mithen and Black 2011), dams were constructed much later (Helms 1981; Whitehead et al. 2008). The ancient Indus civilisation of Harappa is dated to a slightly later period, around 2600–1900 BC. Little direct archaeological evidence of hydraulic systems has been recovered to date (Kenoyer 2000). Overall, with the emergence of complex societies, most ancient cultures began constructing dams, reservoirs, and canals to prevent floods and facilitate transport and irrigation. In comparison, the hydraulic system near the Liangzhu site cluster was constructed very early on, built on a grand scale and boasts a perfect structure. In terms of construction materials, while the ancient Egyptians used blocks of stone to construct their dams and the Harappans used mud bricks or fired bricks to build river dykes, the Liangzhu Culture invented the distinctive use of clay wrapped up with grasses as the main construction material.

2.3.5 Altar and elite cemetery

Altars and elite cemeteries such as Yaoshan and Huiguanshan are located outside the city wall of the Liangzhu City.

Yaoshan is a natural mountain, 35 m above sea level, located 5 km north-east of the Liangzhu City. The Liangzhu altar was first discovered on top of Yaoshan in 1987 (Zhejiang 2003). There are slope protection structures made of stone on the west and north sides of the altar. The top of the altar is flat. The construction procedures of the platform were as follows: Trenches were first dug out and the earth was compacted to form a concentric square structure with a red earthen platform inside, a square of trenches filled with greyish earth in the middle, and a gravel platform on the outside. There are 13 elaborate Liangzhu burials distributed in two rows south of the altar with their graves cutting into the platform (Figure 2.17).

Huiguanshan is a natural mountain about 22 m above sea level, located 2 km west of the Liangzhu City. The altar is similar to the one on Yaoshan and was discovered during excavation in 1991 (Zhejiang and Yuhang 1997, 2001). The altar on Huiguanshan was also dug and built using the natural typology of the hill. The overall shape is rectangular with two steps on the east and west ends. There are two small grooves running north to south on each set of the steps. A similar method of digging trenches and compacting earth, as seen at Yaoshan, was used to make a grey earthen square on the west side of the altar. Four elaborate Liangzhu burials have been excavated at the south-west side of the altar on Huiguanshan.

After many years of researching the artificially constructed altars on Yaoshan and Huiguanshan, Liu Bin made the astonishing discovery that the directions of sunrise match with the four corners of the altar. The south-east corner of both altars points in the direction of sunrise on the day of the winter solstice, at about 135° east of north, while the south-west corner points in the direction of sunset, at about 225°. The north-east corner of both altars points in the direction of sunrise on summer solstice, at about 45° east of north, while the north-west corner points in the direction of sunset, at about 305°. The sun rises due east of the altars on spring and autumn solstices, at about 90° east of north and sets due west of the altars at about 270°. Thus, the altars were most likely used for dating as the length of a year can be accurately determined by observing the sun (Liu 2006).

2.3.5.1 The surrounding area of the Liangzhu City

There are two areas outside the city centre of the Liangzhu site cluster where sites are densely concentrated, one is centred around the Xunshan

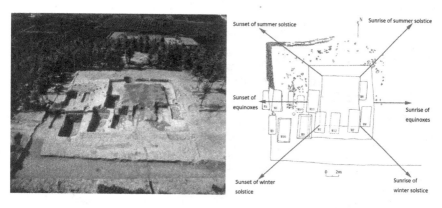

Figure 2.17 Altar and elite cemetery on Yaoshan and diagram of astronomical
observations on the altar

Mountain near Liangzhu Town, the other is in front of the Dazheshan
Mountain, north-east of the Liangzhu City. There are nearly 30 sites in
the group centred around the Xunshan Mountain near Liangzhu Town,
among which the Miaoqian site is excavated in the most detail. Six
seasons of excavations in 1988–1989, 1990, 1992, 1993, 1999–2000,
and 2001 have covered an area of 0.39 ha and found many burials,
houses, ditches and canals, wells, and storage pits and ash pits allowing
us to see a rural settlement in the area surrounding the Liangzhu City.
The area in front of the Dazheshan Mountain, north-east of the Liangzhu
site cluster, is an area where sites were most densely distributed outside
the city centre. So far, 40 sites have been found in this surrounding area
of the Liangzhu City, including the Tangshan site that has a jade workshop and cemeteries in addition to being part of the hydraulic system,
the Yaoshan Altar and the elite cemeteries just mentioned, as well as
associated sites or cemeteries of a lower status like Meijiali, Meiyuanli,
Guanzhuang, and Yaojiadun.

There are many other Liangzhu Culture sites distributed throughout
the 1,000 km² occupied area by the C-shaped basin, some of which were
clustered in groups (see Figure 2.1). For example, 20 Liangzhu sites have
been found near Linping about 30 km from the Liangzhu City. This is
called the Linping site cluster (Zhao 2012). Among sites discovered in
recent years, the Maoshan site is a settlement that is typically situated
on the foothill between a mountain and a river. The site has several well-
defined functional areas, including Liangzhu-period rice paddies, cemeteries, and residential areas (Figure 2.18) (Ding and Zheng 2010). The
Yujiashan site is another important site in this cluster. It occupies an area
of approximately 15 ha and comprises six moated zones (Figure 2.19)
(Lou et al. 2012).

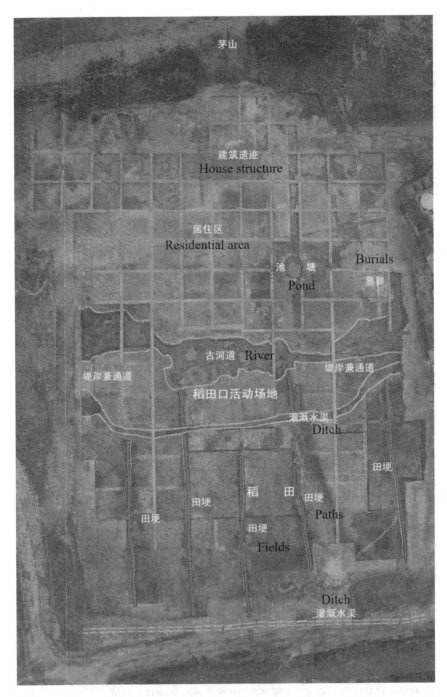

茅山

建筑遗迹
House structure

居住区
Residential area

池　塘　Burials
Pond 墓葬

古河道　River

堤岸兼通道　　　　　　　堤岸兼通道

稻田口活动场地

灌溉水渠
Ditch

田埂

稻　　田
田埂
田埂　　Paths
田埂

田埂
Fields

Ditch
灌溉水渠

Figure 2.18 Top view of the settlement and paddy fields uncovered from the Maoshan site

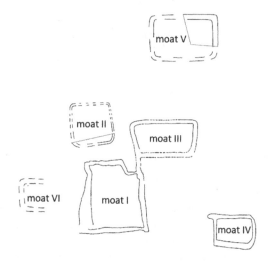

Figure 2.19 Plan of Liangzhu period moats at the Yujiashan site

2.4 Characteristics of the Liangzhu City and Liangzhu Culture

We can summarise the following characteristics about the Liangzhu City and the Liangzhu Culture based upon our excavations and research.

First, the Liangzhu City was one of the largest urban centres among contemporary ones in China and worldwide. It was large in scale and reasonably planned, with a complete structure and a clear layout.

The Mojiaoshan platform is at the centre of the Liangzhu City, surrounded by the inner and outer walls. The Mojiaoshan Palatial Compound occupies 30 ha, the area enclosed by the inner city wall is 300 ha, and the outer circle developed during the last phase of the middle to late Liangzhu period (2800–2300 BC) reached 800 ha.

In addition to the urban structure, there was also a large-scale hydraulic system, altars and elite cemeteries, and expansive rural areas surrounding this urban centre. The hydraulic system and the urban centre comprise the central parts of the planned zone. The area protected by the hydraulic system, where the sites are most densely distributed (east of the Ganggongling dam, south of the Dazheshan Mountain, west of the eastern edge of the site cluster, and north of the southern slopes of the Daxiongshan Mountain), reaches nearly 100 km². The harmony between the layout of the Liangzhu site cluster and related sites and the mountains and streams in the landscape demonstrates the remarkably insightful planning by the Liangzhu inhabitants.

The Mojiaoshan Palatial Compound, which included platforms within the inner city wall, the inner city wall itself, platforms in the area between the inner and outer city walls, the hydraulic system, and some sites in the

Table 2.1 Construction works for the Liangzhu site cluster (unit: 10,000 m³)

Structure	Volume of earth	Volume of rubble	Total volume	Total
Mojiaoshan	228		228	1,005
City wall	110	approx. 10	120	
Platforms inside the city wall	281		281	
Outer circle	88		88	
Hydraulic system	288		288	

surrounding area were all artificially constructed with earth, making the Liangzhu Culture a typical 'earthen civilisation'. The scale of engineering and construction was immense. The amount of earth and rubble used to construct the entire urban structure and the hydraulic system is estimated to be 10,050,000 m³. Even if the outer circle constructed during the middle to late Liangzhu period is not included, the amount of earth used to build the urban centre and the hydraulic system around 3000 BC reaches 9,170,000 m³. This was undoubtedly one of the most difficult and labour-intensive construction projects undertaken worldwide at this time. There is no reliable evidence about the architecture on top of the platforms, so the amount of labour required for logging, transporting and processing timber, and constructing buildings cannot be estimated. For a conservative estimation of 9,170,000 m³ of earth and rubble, it would require 10,000 people working 365 days in a year in construction, and assuming that it took three people to complete one square meter in a day, it would take 7.5 years of continuous construction to finish the project (Table 2.1). In reality, however, based on the agricultural productivity at the time, the logistic stress of continuously applying a large amount of labour to work exclusively on urban construction would have been unimaginable. A more likely scenario would be to employ the traditional method of constructing hydraulic projects in ancient China, which is to operate only during the non-agricultural season in winter and spring; this method ensures that the tools and food supply required for each labourer can be obtained by themselves during the agricultural season with no extra reserve needed. Also, construction can hardly be carried out during the rainy season. Therefore, if workers participated in construction during the non-agricultural season for 100 days a year, it would take 10,000 people 27.5 years to complete all the earthen constructions. Therefore, the construction of the Liangzhu City was an immense project that took decades to complete.

We can make comparisons with urban centres in other ancient civilisations elsewhere in the world. The Sumerian city of Uruk was the largest site during the Ubaid period (fifth millennia BC), with a settled area reaching around 250 ha. It then further expanded into an area of 600 ha during the Late

Uruk by around 2900 BC (Matthews 2003, p. 110, but cf. Pollock 1999 and Algaze 2009 for more conservative estimations). The Anu District and Eanna District are located in the centre of Uruk, surrounded by residences of officials, craftsmen, and farmers as testified by the large-scale architecture, including palaces and temples, discovered by archaeologists (Nissen 2001). The site is also surrounded by over a hundred small (approximately 7 ha) to medium sites (>7 ha) each (Pollock 1999, also cf. Yang 2014).

The ancient Egyptian city of Memphis was the capital during the Old Kingdom period (2650–2175 BC), possibly also as early as the Early Dynastic period (c. 3000–2650 BC), contemporary to the early to middle Liangzhu period (c. 3000–2700 BC). However, overlain by deep (c. 5 m) alluvial deposits, its size and layout has not been fully understood due to the lack of archaeological surveys (Giddy and Jeffreys 1991; Jeffreys and Trvares 1994; but cf. Bunbury for some new survey data, 2018).

The largest Harappan site is Mohenjo Daro, which occupies an area of 2.5 km^2. Its most distinctive characteristics include meticulous urban planning and labour-intensive construction. The site is composed of the Lower City and the Citadel. The Lower City has many criss-crossed streets oriented east-west and north-south dividing the urban space into rectilinear grids, in which there are many houses and wells. Important buildings such as public baths, granaries, assembly halls, and possibly defensive structures are found inside the Citadel (Kenoyer 2000; Kenoyer and Heuston 2005).

This comparison shows that the scale and overall structure of the Liangzhu City are of the same level to the well-studied late-Neolithic urban centres listed in the preceding text.

Second, archaeological features of the Liangzhu City and research on jade objects show that divine power and rulership were closely connected in the Liangzhu Culture. Thearchy and kingship had such a prominent position in Liangzhu society and the insignia was widespread across the Taihu Lake region through time (cf. Chapters 3 and 4). More importantly, the designs of the insignia were more or less the same, promoting some to suggest that they represent the 'monotheism' of Liangzhu society (see Chapter 5 and Zhang 1997; Liu 2007; Fang 2015). The jade *yue* axe, as a symbol of kingship, occurred commonly in Liangzhu elite tombs, was used like a sceptre (Liu 2007). In other words, the Liangzhu Culture was a civilisation bound by divine power.

According to the layout of the Liangzhu City, the Mojiaoshan Palatial Compound is located right in the centre of the walled city, while the ruler's possible residence and the most important palace structure was located on top of Damojiaoshan, overlooking the city as well as the surrounding landscape stretching from the Dazheshan Mountain to the Daxiongshan Mountain. This location gave the ruler a broad commanding perspective that reflected the prestige of the rulership. This is the perfect example of rulers residing in a central and elevated location. The placement of the palace district on the high ground within the city is comparable to the site of many capital cities during the historic periods in China. Judging from the arrangement of burials at the Fanshan and Yaoshan sites, burials with the largest amount of funerary goods are always placed in the middle of the rows of tombs (Zhejiang 2003, 2005a), which also reflects the practice of placing the elite burials in a prime location of the cemeteries.

Liangzhu jade objects illustrate the close relationship between rulership and divine power in Liangzhu society. To date, Yaoshan and Fanshan are the sites where the most abundant jade objects had been excavated. The Liangzhu Culture created a ritual jade system typified by *cong* tubes, *bi* disks, *yue* axes, *guanzhuangshi* cockscomb-shaped objects, *sanchaxingqi* three-pronged objects, *huang* semicircular ornament, and *zhuixingqi* awl-shaped objects. The *shenhui* insignia motif was not only carved on many jade objects but also directly influenced the shape of ritual jade objects like *cong* tubes, *guanzhuangshi* cockscomb-shaped objects, and ornaments of *yue* axes. The system of ritual jade and the use of the *shenhui* insignia on Liangzhu jade across the entire Taihu Lake region was very consistent. Clearly, they were means and messages for maintaining the sociopolitical control in Liangzhu society. There was strong social cohesion in the Liangzhu Culture as well as a unanimous divine worship. Compared to jade objects from the Songze Culture, the quantity, size, type, and craftsmanship of Liangzhu jade objects had clearly undergone great development, which was almost an instantaneous phenomenon (see Chapter 5 for a discussion on the relationship between Songze and Liangzhu). This kind of leaping development was perhaps a phenomenon that emerged alongside the rise of kingship, as contested by the construction of large infrastructural facilities (the city, the hydraulic system, and so forth) and the unprecedented focus on fine craftsmanship and motif system as well as the underlying ritual hierarchy. The Liangzhu people designed a set of ritual jade to reflect identity and status based on divine, royal, and military power. While rituals were systematised, the depiction of the *shenhui* insignia was consistent.

Take *cong* tubes and *yue* axes as examples. All characteristic Liangzhu *cong* tubes are carved with the *shenhui* insignia. In fact, to depict this sacred image was the *raison d'être* of the Liangzhu *cong* jade tube. It was probably a ritual object used for religious-like ceremonies and was exclusively owned by a small number of priests and rulers. Rulers might have been priests themselves (Figure 2.20). As for *yue* axes, Lin Yun pointed out in his article 'Discussing the Word "King" (*Shuo Wang*)' that the character 'king' 王 in oracle bone script was created based on the shape of an axe (Lin 1965). Therefore, the *yue* axe was an important symbol of royal and military power. A pair of symmetrical *shenhui* insignia motifs carved on the *yue* axe excavated from tomb no. M12 at the Fanshan site, the most elaborate Liangzhu burial, demonstrate the close connection between divine power and kingship (see Chapter 3 for a detailed analysis of the meaning of the motifs). The shape of *duanshi* end ornaments of jade axes borrows the shape of a cockscomb-shaped ornament folded in half. If the *guanzhuangshi*

Figure 2.20 Shenhui insignia carved on the 'king' of *cong* tube, from the Fanshan
site tomb no. M12

cockscomb-shaped object can be seen as a sacred crown, then the *duanshi*
end ornament of axes also represents a folded sacred crown in terms of
its shape as half profile of the *guanzhuangshi*. Putting a jade ornament
that represents a sacred crown on top of a *yue* axe represents the idea that
the right of kings is endowed by the gods. The syncretism between royal
power and divine power elevates the *yue* axe above a common weapon and
transforms it into a kind of sceptre. By wearing a sacred crown, the jade axe
becomes a symbol of the divine will (Figure 2.21) (Liu 2013).

Jade objects represent the supreme authority of divine power and the
syncretism between divine power and kingship, showing that the Liangzhu
civilisation used divine power to form and consolidate social cohesion. This
is in stark contrast with the Xia Culture (c.1900 BC onwards) represented by
Erlitou (cf. Chapter 6 and Xu 2009) but is very similar to Egyptian dynasties
(Baines 1995, 2014).

Third, a 'pyramid-like' social hierarchy emerged as power and wealth
became concentrated in the hands of the few in Liangzhu society, leading
to social stratification. A remarkable multi-tier system among the Liangzhu
settlements was created. There were mega-scale urban centres like the
Liangzhu City as well as small to medium sites that only occupied a few
thousand or tens of thousands of square metres in area.

The rich Liangzhu funerary goods also allow us to see the degree of
social differentiation in Liangzhu society. Zhang Zhongpei pointed out in
his article 'Liangzhu Cemeteries and the Complex Society They Represent'
that Liangzhu cemeteries can be divided into six ranks. The first rank is
represented by the Yaoshan cemetery. Funerary goods include *cong* tubes
and *bi* disks, and tomb occupants belonged to the elite class that controlled military and royal power. The second rank is represented by burials
belonging to the three culture phases of the Fuquanshan site. Funerary

Figure 2.21 The 'king' of *yue* axes, from the Fanshan site tomb no. M12

goods include *cong* tubs, *yue* axes, and stone axes. Tomb occupants included soldier-labourers in addition to people possessing divine and military power. The third rank is represented by burials belonging to the second culture phase of the Fuquanshan site. Funerary goods include *yue* axes and stone axes but not *cong* tubes. Tomb occupants were people with military power as well as soldier-labourers. The fourth and fifth ranks are represented by burials from the first cultural phase at the Fuquanshan site and tombs no. M3, M8, M9, and M10 at the Maqiao site. Funerary goods include stone axes and show that the tomb occupants were soldiers. The presence or lack of jade objects determines whether the burial belongs to the fourth or the fifth rank. The sixth rank is represented by tombs no. M4, M5, M6, and M7 at the Maqiao site where only a small amount of pottery and no funerary goods were found. Tomb occupants were from the poorest class in society (Zhang 2012).

Fourth, advanced agriculture and craftsmanship were the preconditions for the emergence of the Liangzhu Culture and the construction of the Liangzhu City.

Around 25,000–60,000 people lived in the Sumerian city of Uruk that occupied an area of 250–600 ha (if taking the conservative estimation of

Figure 2.22 Carbonised rice recovered by flotation, found in pit no. H11 on the east slope of Mojiaoshan

population density of 100 persons per hectare, see Pollock 1999, p. 65), while Mohenjo Daro occupied a similarly sized area as Uruk and likely had a population size of about 30,000 to 40,000. Based on these numbers, the population residing in the 800-ha Liangzhu site cluster may have been larger than these two sites.

A large population must be supported by high agricultural yields. A large pit, no. H11, was discovered during the excavation of the east slope of the Mojiaoshan platform during the 2010–2012 excavation season. The pit has an irregular oval shape, with a longer diameter of 29.3 m and a shorter diameter of 17.7 m. The deposits inside the pit can be divided into three layers. The first and third layers are composed of dark greyish soil with abundant charcoal, carbonised rice, lumps of burnt bricks, plant ash, and a small amount of straw ropes. Archaeobotanical analysis estimated about 13,000 kg of rice remains from this pit fill. Given that the deposits were mainly composed of domestic rubbish, which normally does not contain such a large proportion of edible rice, the disposal of such a large amount of carbonised rice by the Liangzhu inhabitants seems unusual. This was likely where burnt rice and ashes were disposed of after granaries caught on fire. The layers in the pit revealed that these types of fires likely happened twice (Figure 2.22). A considerable amount of carbonised rice was also discovered while excavating the south-west slope of Mojiaoshan in 2013.

These discoveries have made us realise that there was a large reserve of rice in the palace area where large granaries were located.

While excavating Meirendi and Mojiaoshan, we also carried out coring surveys since 2010 to search for Liangzhu-period rice paddies around the Liangzhu City and nearby. However, no trace of rice paddies has been found yet. This shows that most likely no rice was produced inside the enclosed areas of the Linagzhu centre. Instead, rice was supplied by the surrounding area of the Liangzhu City and by settlements outside the Liangzhu site cluster. Therefore, there should have been a system of tribute payment in place.

The discovery of rice paddies at the Maoshan site in the Linping site cluster offers important archaeological evidence on agricultural production in small villages during the Liangzhu period. Maoshan is a typical site located on a foothill with residential areas, cemeteries, and rice paddies discovered. Rice paddies are located on the low ground south of the foothill of the Maoshao

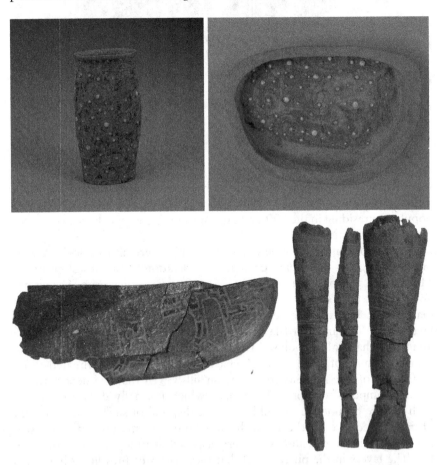

Figure 2.23 Liangzhu lacquerware. 2.23a: repaired lacquerware from the Fanshan site; 2.23b: lacquerware from the Bianjiashan site

Figure 2.24 Incised pottery

Mountains. Rice paddies of the middle Liangzhu period were made of strips of fields that were not very large, with each paddy ranging from 1 or 2 m² to 30 or 40 m². These small paddies developed into immense paddies occupying 5.5 ha during the late Liangzhu period. Five north-south–oriented fired earthen paths and two east-west–oriented canals have been discovered. These paths are 17 m to 19 m long, dividing the entire area into large paddies about 1,000 m² to 2,000 m² in size (see Figure 2.18) (Wang 2013).

Supported by this highly developed agricultural production, division of labour became highly organised as craftsmen were freed from agricultural labour and able to specialise in the production of jade, lacquerware, pottery, and textiles. These craftsmen created rich and diverse forms of art including many masterpieces of fine jade, lacquerware (Figure 2.23), and incised pottery (Figure 2.24). Some fine crafts were monopolised by the elite and became state-controlled crafts, such as jade production.

2.5 Concluding thoughts

Archaeological discoveries at the Liangzhu City have gained full appreciation within Chinese archaeological circles. The prominent archaeologist, Yan Wen Ming, once stated,

> The Liangzhu City is the largest prehistoric [Chinese] urban site discovered to date. It is the only site where we have a clear understanding of the expansive city and its surrounding areas. The Liangzhu City and surrounding areas were planned and have clear functional divisions. Mojiaoshan platform is the core at the Liangzhu City and the political centre of the Liangzhu Culture. The capital of the Liangzhu Culture is right here.

Zhang Zhongpei, another well-known archaeologist, also stated after several visits to the site, 'The Liangzhu City is unique in China. Its meaning and significance is comparable to Yinxu. As the largest urban site of its period in China, it can be called the "first and greatest city in China."'

Important parts of the Liangzhu City including Fanshan, Yaoshan, Huiguanshan, and Mojiaoshan as discussed previously are among the ten greatest archaeological discoveries between 1986 and 1990 or among the ten greatest archaeological discoveries of their year. The excavation of the Liangzhu City was also listed among the ten greatest national archaeological discoveries of 2007, awarded second place for the Field Archaeology Prize awarded by the State Administration of Cultural Heritage of China during the 2007–2008 season and first place for the 2009–2010 season. New discoveries at the Liangzhu City have also been listed among the ten greatest archaeological discoveries worldwide in 2013 at Shanghai Forum. More than 70 archaeologists from 28 countries listened to a presentation about recent discoveries at the Liangzhu City during the Shanghai Archaeology Forum in 2013 and visited the city wall, the Mojiaoshan platform, and the Liangzhu Museum after the conference. This boosted the worldwide influence and fame of the Liangzhu City. After being listed on the tentative list of UNESCO World Heritage Sites in 1994, 2006, and 2012, preparations are now complete for nominating the site as a World Heritage Site for 2019. The Liangzhu City is now moving beyond its importance in Chinese archaeology and towards international recognition for its significance in ancient civilisations developments in urban planning.

3 Power and belief

Reading the Liangzhu jade and society

Qin Ling, translation by Tang Xiaojia

The rise of the Liangzhu Culture (3300–2300 BC) plays a distinctive role in China's prehistoric cultural development and, of course, early Chinese civilisation. Compared with other contemporary archaeological cultures, jade is the most prominent material recovered from Liangzhu. It is closely related to the settlement hierarchy in Liangzhu society, carries important messages of cultural transmission processes beyond the Liangzhu area, and, more significantly, it is the major indicator of the Liangzhu belief system and spiritual worldview.

The central position of the Liangzhu site cluster in the Liangzhu Culture was first recognised with the discovery of the Fanshan and Yaoshan cemeteries in 1986. Our understanding of the central importance of this site cluster has been enhanced by a series of important discoveries since 2000. Due to the limitations of archaeological materials, most investigations on ancient societies have started by a central focus on objects to reconstruct local histories. For Liangzhu society, the most prominent position was, in no doubt, occupied by jade and the people who produced and used jade artefacts. Therefore, we begin this chapter with a closer look at a 'king's' burial, and through the interpretation of its grave goods we hope to draw a detailed understanding of the two key concepts of Liangzhu society: Power and belief.

3.1 Funeral of the 'king': The Fanshan tomb no. M20

One of the major difficulties in archaeological analysis of ancient burials is how to differentiate between funerary objects by examining the burial goods: Which were made specifically for funerary purposes and which had been used before death and shared by the community? Which artefacts can be taken as status symbols (materiality of social power), and which were reflections of communal cultural practices and people's spiritual world, in other words, an indication of a culture's belief systems?

At Liangzhu, we are fortunate to have the Fanshan tomb no. M20 as an example for discussion. Fanshan is clearly a 'king-level' cemetery in

Liangzhu and M20 is one of the most important tombs at Fanshan as it yielded the largest number, richest category of jades among the cemetery, as well as the most sophisticated craftsmanship of jade, right after those found inside burial no. M12 at the same cemetery. A piece-by-piece examination of the grave objects found within M20 is the best way to reconstruct the burial practices that took place during the funeral for the tomb occupant (the 'king') of M20.

The Fanshan cemetery is located in the north-west corner of the Mojiaoshan platform in the Liangzhu site cluster. It is 5 to 6 m above ground and surrounded by several ponds of various sizes, which may have been dug to provide soil for the construction of the Fanshan platform. The platform is currently 90 m wide, east-west and 30 m wide, south-north, with an overall size of 2,700 m². During the 1986 excavation, 11 Liangzhu-period graves were found in the western half of the Fanshan platform (Figure 3.1) (Zhejiang 2005a).

Of the 11 tombs, tomb no. M21 was partly destroyed. Archaeologists were only able to collect a few objects from the disturbed soil and one tall *cong* tube is typical of the late Liangzhu period style. Another tomb, no. M19, was also not well preserved, without a clear grave-pit identified. Therefore, in archaeological research, these two tombs are normally excluded and the remaining nine burials compose the main part of the Fanshan cemetery. All nine burials are aligned in an orderly manner and dated to the early phase of the middle Liangzhu period (ca. 3000–2800 BC).

Figure 3.1 Distribution of tombs at the Fanshan cemetery (from south to north)

We can see that there are two rows of tombs distributed from north-west to south-east at the Fanshan cemetery. In the northern row, from west to east, there are tombs no. M18, M20, M22, and M23; and in the southern row, there are tombs no. M15, M16, M12, M17, and M14 (Figure 3.1).

M20 is located in the middle position of the northern row, with a large burial pit, measuring 4 m in length, 2 m in the southern end and 1.75 m in the northern side. The remaining depth is roughly 1.32 m deep. Archaeologists discovered 538 grave goods in the tomb, including 499 jade items, 24 stone artefacts, 2 pottery vessels, and some ivory objects and shark teeth. The burial goods found within the tomb and within the set of coffins have different spatial arrangements. The spatial arrangements of burial objects both in and outside the burial equipment are clear. Even though the human skeleton is not preserved due to the acidic soil conditions, from the burial artefact assemblage, it can be argued that the tomb occupant of M20 is male. The funerary practices had both inherited past traditions from earlier tombs and developed an unprecedented middle-Liangzhu-period characteristics, through which we are able to get a comprehensive understanding of how a 'king-level' individual was represented in the material culture of Liangzhu (Figure 3.2).

Figure 3.2 Fanshan tomb no. M20 (from south to north)

Figure 3.3 Cong-style *zhuxingqi* cylindrical objects from Fanshan tomb no. M20

First, the position of the artefacts that were placed outside the coffin of M20 can be taken as representative of the highest-ranking individual at the Liangzhu site cluster. Three *cong*-style *zhuxingqi* cylindrical objects (Figure 3.3) were found on the top, middle, and bottom parts of the coffin. One of these *zhuxingqi* cylindrical objects is a semi-finished product, on which a rough blueprint of the human face pattern was achieved but had not been finely processed; only the nose and the surrounding lines were carved and grounded. This semi-finished piece tells us that some of the burial goods were not owned or used by the tomb occupant before their death but made in a rush specifically for burial. We can also see that within the Liangzhu elite community, relatively fixed burial and funerary practices had been formed. It is because of this rather well-acknowledged implementation of funerary practices that semi-finished items of specific shapes could be used and appeared at certain positions within the 'king's' tomb. Compared with other cemeteries of the Liangzhu Culture, it is almost certain that being buried with a set of *zhuxingqi* cylindrical objects was typical among privileged male burials.

At the northern (bottom) end of the coffin were one ivory item, two jade bi disks, and a necklace made of seven jade tubes. The bi disks were placed close together with the jade tubes, under which is the shape of a vermillion-colour container, possibly a scarlet-lacquered vessel. Due the poor preservation condition, the excavators failed to collect the fragments of lacquer objects. However, what we can say with certainty is that these jade items were originally placed upon a big lacquer plate which was then put outside the coffin in the burial.

We shall start our discussion on the objects inside the coffin from the southern end (near the head). It is a normal practice for high-class males of Liangzhu Culture to have a *guanzhuangshi* cockscomb-shaped object, a *sanchaxingqi* three-pronged object, and a set of *zhuixingqi* awl-shaped objects; within Liangzhu site cluster, an additional set of four semicircular ornaments would be used as well. In tomb no. M20, archaeologists have found all these fixed sets of ornaments. If one wants to order the symbolic importance of these various jades, which might be directly related to social power, the set of semicircular ornaments ranks first due to its exclusive use in Liangzhu site cluster, followed by set of *zhuixingqi* awl-shaped objects, *sanchaxingqi* three-pronged objects, and *guanzhuangshi* cockscomb-shaped object.

These jade semicircular ornaments (Figure 3.4) so far have only been found from the Fanshan and Yaoshan cemeteries. Considering the hole opening at the back of these jade ornaments, they might be used as pendants attached to crowns. However, the number of tombs, which have yielded these semicircular ornaments, is quite limited, and we are uncertain as to what these crowns were used for and their symbolic meanings.

Sets of *zhuixingqi* awl-shaped objects are considered to be specifically related to male identity in Liangzhu society. Multiple pendants in odd numbers are normally found near the heads of the deceased in Liangzhu elite and non-elite burials. The number of *zhuixingqi* awl-shaped objects we have from Liangzhu range from 3 to 11 only in odd numbers in male burials. It would seem that these *zhuixingqi* awl-shaped objects were not exclusive to high-status members, as they are also present in ordinary graves, suggesting that the jade resource was available to 'ordinary' people. On the bottom of each *zhuixingqi* awl-shaped object were tenons and drilled holes, indicating that they were originally attached to some sort of organic material

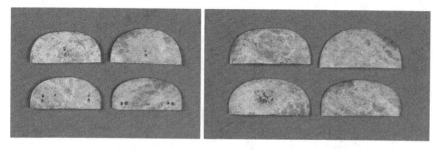

Figure 3.4 A set of semi-circular ornaments from Fanshan tomb no. M20 (back and front sides)

with the pointed head upwards. The set of *zhuixingqi* awl-shaped objects reflects both the power and belief in Liangzhu society. On the one hand, appearance or absence of carved motifs and the number of *zhuixingqi* awl-shaped objects in one set clearly indicate different social ranks; on the other hand, they were only found in fixed positions, using certain methods and in male graves, which shows that they must have played some specific roles during funerary rituals. For example, in M20, archaeologists have found a set of nine, with eight identical pieces of plain *zhuixingqi* awl-shaped objects and only one longest carved with patterns (Figure 3.5). Thus, as a whole, they showed the special elite status of the occupant.

Sanchaxingqi three-pronged objects are another type of male-only ornaments, which were often placed together with a long jade tubes on the heads of the decreased. Based upon the positions of the small holes, it can be inferred that the *sanchaxingqi* three-pronged object was fixed to something to be used together. The one found in M20 was in semi-disk shape. The three prongs were similar in length and on the back side there was a bump on which tiny holes were drilled (Figure 3.6a). This indicates that there were different purposes for the obverse and reverse sides of the *sanchaxingqi*

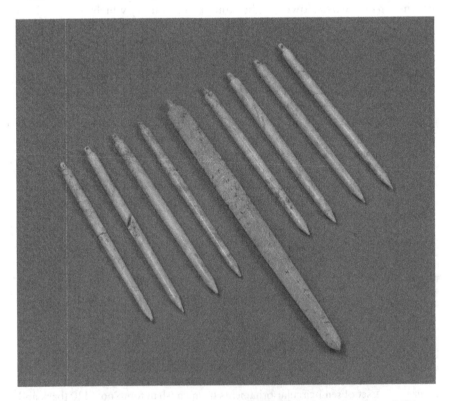

Figure 3.5 A set of *zhuixingqi* awl-shaped objects from Fanshan tomb no. M20

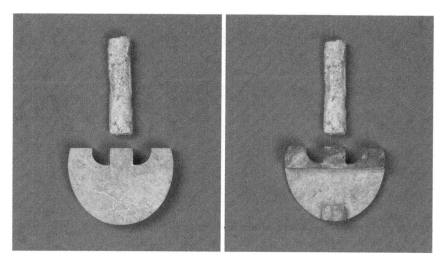

Figure 3.6a Sanchaxingqi three-pronged object from Fanshan tomb no. M20 (front and back view)

three-pronged objects. To some extent, if comparing the one from M20 with another *sanchaxingqi* three-pronged object from tomb no. M14 at Fanshan, we can see that there were detailed carvings on the reversed side of the M14 jade as well (Figure 3.6b). This shows that both sides of *sanchaxingqi* three-pronged objects could be used for display. Another more commonly found style of three-pronged objects was the ε epsilon (or 山 in Chinese character) shape, which was found in various status burials even in Jiaxing (north Zhejiang) area. Considering that they were sometimes carved on both sides (Figure 3.7), it is possible that they were used for display purposes from multiple angles. Therefore, we may argue that these *sanchaxingqi* three-pronged objects were used in daily life or rituals and not specially produced for burials or funerary rituals.

However, the practice of burying *sanchaxingqi* three-pronged objects does not appear in the wider Liangzhu region. It is only seen at the Liangzhu site cluster and nearby areas, such as in Jiaxing in northern Zhejiang province. No high-status graves were buried with *sanchaxingqi* three-pronged objects in the areas to the east and north of Taihu Lake (see also Chapter 4).

Another unique type of Liangzhu jade items that are normally found decorating the heads of the deceased are the *guanzhuangshi* cockscomb-shaped objects. Due to a complete ivory comb with a *guanzhuangshi* cockscomb-shaped object excavated from tomb no. M30 at the Zhoujiabang site in Haiyan, the function and usage of these ornaments became clear to archaeologists. Newly published archaeological reports name these *guanzhuangshi*

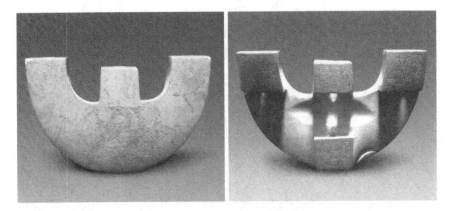

Figure 3.6b Sanchaxingqi three-pronged object from Fanshan tomb no. M14

Figure 3.7 Sanchaxingqi three-pronged object from Yaoshan tomb no. M7

cockscomb-shaped objects 'jade comb head'. As a special type of hair accessory, the jade comb head may not have been a very strong indication of social
power. First, almost in every Liangzhu burial, there is one and only one jade
comb head, which clearly was not gender-specific. Second, jade comb heads
appear within the most modest Liangzhu graves, which means that they were
not exclusively associated with the high elite class. However, the key elements
of the Liangzhu belief system were embodied in the shape and design of the
jade comb head. Jade comb heads, unearthed from the Fanshan cemetery,

were the most typical representations: They are upside-down trapezoids with notch (介-shaped) in the middle part, under which were oval-shaped small holes. The *guanzhuangshi* cockscomb-shaped object (jade comb head) found from M20 was the only one currently known without such a cutout hole but left with multiple string-cutting traces on the surface. It may be that this jade comb head was not used before death but was specially made for ritual purposes during the funeral. To understand the real meanings behind the shape of these Liangzhu jade comb heads, the carving patterns and other artefacts decorated with similar patterns need to be referenced (Figure 3.28).

Cong tube, *bi* disk, and *yue* axe were the three most important forms of jade objects in the Liangzhu Culture. Fanshan tomb no. M20 in total yielded 4 *cong* tubes, 43 *bi* disks, and 25 *yue* axes (1 jade and 24 stones). A detail analysis of the M20 jade assemblage of *cong*, *bi*, and *yue* provides us with an unusual glimpse into this typical 'king's' tomb.

Three *cong* tubes were found near the right arm, on which a stone *yue* axe was placed. Another *cong* tube was uncovered near the left arm, which was throughed with an ivory item. This set of ivory in *cong* tube was positioned shallower than other items inside of coffin and thus should be treated differently when discussing its function. Underneath this special *cong* tube, there was the *yue* axe, which also demonstrates that the original position of ivory in *cong* tube assemblage should be different from other *cong* tubes.

Among the three *cong* tubes found on the right-hand side, item no. M20:122 is assumed to be the rarest one. Upon closer examination, the patterns on this *cong* tube were comprised of double horizontal lines zone with the sacred human and another set of double horizontal lines zone with animal-face motif (Figure 3.8a). In general, the double horizontal lines have

Figure 3.8a *Cong* tube, Fanshan M20:122

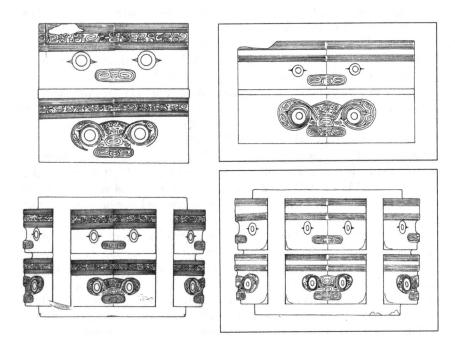

Figure 3.8b Comparison of the Fanshan and Yaoshan *cong* tubes –left: Fanshan M20:122; right top: comparison of decoration style, Yaoshan M12 (2784); right bottom: comparison of composition, Yaoshan M12 (2786)

been regarded as a variant of the headdress of the sacred human. Only when the animal face appeared alone will there be double horizontal lines zone decorating the upper section of the *cong* tube. Therefore, cases like the M20:122 jade *cong* tube are extremely rare. The only other known example is a *cong* tube collected from the Yaoshan cemetery, tomb no. M12 (Figure 3.8b bottom right). In addition, a similar design of the sacred human with animal-face motif and the horizontal lines can be found on the jade *cong* tube no. M12:2789 at Yaoshan. This design was adopted to better fit the upside-down trapezoid-shaped face of the sacred human (Figure 3.8c).

Diving deeper into the details of the *cong* patterns, it seems that item no. M20: 122 was chronologically earlier than the other burial objects in M20, which might belong to the early collection of the tomb occupant or his ancestors. The double-circled animal eyes drilled on the face pattern are bigger than ordinary ones, and the eye corners are depicted with double lines. The space between the double lines was filled with curly patterns and the eyelids were also carefully filled with carved pattern. Similar decorative patterns are also seen from the *cong* tube from Yaoshan tomb no. M12 (Figure 3.8b top right). With these prominent similarities, we can see that the tomb occupant of Fanshan tomb no. M20 was closely associated with

Figure 3.8c Cong tube, Yaoshan M12 (2789)

the individual buried in Yaoshan tomb no. M12. The shared belief system and the detailed jade decorative repertoire indicate that the high elites of different subgroups at Liangzhu site cluster were connected within the same jade production and allocation network.

The other two jade *cong* tubes excavated from Fanshan tomb no. M20 have a simpler decorative design, which were decorated with double horizontal lines with the sacred human face. Judging from the shape of these two *cong* tubes, we think that they represent two general categories of Liangzhu *cong* tubes, the bracelet-like type and vertical tube-like type. Item no. M20:121 has a round cross-section and, as a bracelet-like type, shows a different shaping technique and procedure from types of tube-shaped *cong* (Figure 3.9a). In the late Liangzhu period, when the tube-style *cong* got bigger and taller, carvings for the human faces became simpler as well. In contrast, round-shaped *cong* tubes as a special kind of jade tube to be worn as a bracelet continued to be produced and used until the late Liangzhu period, for example, in elite burials of the Sidun, Fuquanshan, and Hengshan sites.

Item no. M20:123 at Fanshan, as a single-sectioned cong tube, has two arc-shaped face patterns at the bottom on each side. This was a very rare decoration on Liangzhu jade, with only five examples known so far: Items discovered from the Fanshan tombs no. M12 and M20, the Yaoshan tombs no. M7 and M12, and the Huiguanshan tomb no. M2 (Zhejiang and Yuhang 1997; Zhejiang 2001). It is generally agreed that this arc-shaped face pattern can be compared with the animal-face and dragon patterns of the early Liangzhu jades' decoration. Since the complete sacred human face was usually carved within an upside-down trapezoid shape during middle Liangzhu

Figure 3.9 Comparison of *cong* tubes from a top view–left: Bracelet type, Fanshan M20:121; right: Tube type, Fanshan M20:123

period, the arc-shaped human face reflecting a combination of human and animal should be considered.

Another *cong* tube, item no. M20:124, which was placed in a higher position than other grave objects, was made of four vertical sections and decorated with two sets of the sacred human with animal-face motifs. On the sides of the animal-face patterns, there are bird-shaped decorations, on which one can observe traces of polishing; the noses of the sacred faces were not carefully carved. However, the animal eyes have very sharp angles, an indication of rather later Liangzhu style. On the whole, the completion degree of item no. M20:124 is comparatively poorer than most other burial artefacts in M20 and this *cong* tube is assumed to be specially produced for burial rather than owned by the tomb occupant when he was alive.

In the same way, if we look at the structure of this *cong* tube, we see that the sacred human with animal-face motif is a typical fixed ornamental scheme of the Liangzhu Culture, including the double horizontal lines on top of the human. *Cong* tubes decorated with such patterns were widely distributed in elite burials around Taihu Lake. However, *cong* tubes with a three-section structure (human-animal-human from top to bottom) were

Figure 3.10a Cong tube, Fanshan M20:124

less commonly found, except for a few cases seen at the Fanshan, Yaoshan, Sidun, and Fuquanshan sites. Objects with similar structure to item no. M20:124 (Figure 3.10a), which with four sections of repeated 'line-human-animal' sets are extremely rare, and it is highly likely that they are exclusive to Fanshan tomb no. M12. Among the six cong tubes in this tomb M12, four were decorated with this human-animal-human-animal four sections form, displaying a good consistency (Figure 3.10b). This ornamental scheme might indicate the social status of the tomb's occupant and, in this case, also indicate the inheritance from M12 by M20. The reason why similar cong tubes have so far not been found at other locations could be due to the limitations of jade material or other social restrictions.

From the perspective of decorative patterns, what is most interesting about item no. M20:124 is that it was decorated with bird-shaped patterns. Bird, not at all an ordinary pattern on cong tubes, was used on guanzhuangshi cockscomb-shaped objects (Figure 3.11), sanchaxingqi three-pronged objects, and huang semi-circular ornaments as a side 'help' for the main pattern of human or animal. Apart from item M20:124 (Figure 3.10a), the only other examples of bird pattern on cong tubes were the two objects from Fanshan tombs no. M12 (Figure 3.10b top right and bottom left) and two items from the Fuquanshan and Wujiachang sites (Figure 3.12), respectively (Shanghai 2000). The three pieces from the Fanshan cemetery showed high similarities both in terms of structure and in the arrangements of the bird decorations. The one from the Fuquanshan cemetery has a single set of sacred human with animal-face pattern, with the bird carved on the sides of both human and animal. The Wujiachang cong tube is made of a three-section structure: Human-animal-human from top to bottom and with birds on the sides of animals only, same as those from Fanshan. From all four pieces, we might say that the differences in design and motif on cong tubes are the best

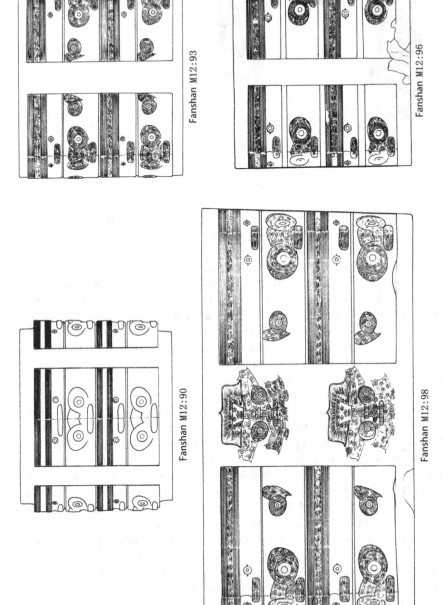

Fanshan M12:93

Fanshan M12:96

Fanshan M12:90

Fanshan M12:98

Figure 3.10b Same structure of four sections form on the *cong* tubes from Fanshan tomb no. M12

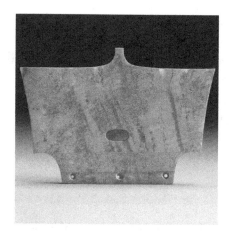

Figure 3.11 Birds on *guanzhuangshi* cockscomb-shaped object from Yaoshan tomb no. M2:1

Figure 3.12 Birds on the *cong* tubes from Fuquanshan tomb no. M9 (left) and Wujiachang tomb no. M204 (right)

indicators of social status in Liangzhu society and the multisection ones show this social hierarchy to an extreme. The incorporation of this bird pattern further reflects the close relationship between the tomb occupants of Fanshao M12 and M20 as well as between the various regional elites social groups of the Ququanshan-Wujiachang site cluster and the Liangzhu site cluster.

Looking further into the carving detail of item no. M20:124, we see that the shape of the eyes was almost identical to those on the Fuquanshan *cong* tube in tomb no. M65 (Shanghai 2000) (Figure 3.13). Also, both items were made from the same kind of raw material with the so-called chicken-bone white colour with its vitreous luster, unique to the area. It can be argued that the Fuquanshan *cong* tube is a finished product acquired from the Liangzhu centre.

Figure 3.13 Animal eyes on *cong* tubes from Fanshan tomb no. M20 (left) and Fuquanshan tomb no. M65 (right)

Taken as a whole, the four *cong* tubes from Fanshan tomb no. M20 provide us with a great deal of information on the occupant and his social connections. Within the Fanshan cemetery, tombs no. M12 and M20 undoubtedly had the closest relationship in time while the burial date of M20 was slightly later than M12. Meanwhile, Fanshan tomb no. M20 shared a similar jade working tradition with Yaoshan tomb no. M12 and other high-status burials in Fuquanshan cemetery.

One jade *yue* axe can be found in every male elite tomb at Liangzhu site cluster, such as at the Fanshan and Yaoshan cemeteries, closely related to ideas of male social status in Liangzhu society. In Fanshan tomb no. M20, archaeologists found a *yue* axe near the left shoulder of the occupant. This axe was placed with a boat-shaped top-end ornament of axe handle. Judging from the position of the end ornament, the blade of the jade axe must have been later disturbed pointing northwards. Compared with other Liangzhu jade types, the production of axes showed more diversity both in terms of the size of the objects and the number and position of drilled holes on each individual axe. The M20 jade axe is long and slim and has a small, single-sided drilled hole near the top while the main hole was made by double-sided drilling. Between the two holes there are surface markings of bindings (Figure 3.14a).

The jade axe as a male status symbol can be traced back to the Late Songze Culture (*c.*3500–3300 BC) in the Taihu Lake area. The Nanhebang cemetery in Jiaxing has yielded such examples (Zhejiang 2005b). It has been assumed that the practice of burying jade axes in male tombs became institutionalised during the transitional period from the Songze to Liangzhu (also see Chapter 5). At the Puanqiao site in northern Zhejiang province, archaeologists found that the practice of every male being buried with one

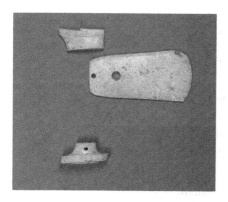

Figure 3.14a Jade *yue* axe and end ornaments from Fanshan tomb no. M20

stone axe was already in place (Qin 2014). When it came to the Yaoshan and Fanshan cemeteries, this practice became truly institutionalised in high elite tombs; jade axes were buried together with other decoration on its handle, including *mao* (top) end ornament, jade *dui* (bottom) end ornament, as well as small inlaid jade pieces (e.g., in Fanshan M14), pendants, and *cong*-style tubes (e.g., in Yaoshan M7). Through the transition from stone axes with holes to jade axes with decorations, axes gradually acquired a crucial position among elite circle in Liangzhu society, which might be associated with monarchical power, unlikely with military power (cf. Chapter 5).

Along with the *yue* axe, 24 pieces of stone axes were also recovered from M20, representing another very distinctive feature of this burial assemblage (Figure 3.2, Figure 3.14b). Only one of the stone axes was buried near the head. The rest were all stacking together with the jade *bi* disks near the lower body or the feet. They were all made of the same raw material and were quite uniform in shape. All of them have a big drilled hole in the middle but none of their blades had been sharpened for real function. Therefore, they were most likely made specially for funerary purposes.

The stone tools industry in the Lower Yangtze was highly developed since the Beiyinyangying-Lingjiatan period (*c.*3600–3300 BC). Adzes, axes with drilled holes, and arrowheads were the most common forms produced. In the early-middle Songze period, such as at Lingjiatan and Dongshancun cemeteries, high-status tombs were buried with large numbers of stone axes and adzes. However, in the south-eastern part of the Taihu Lake, such Ningzhen traditions were not present. Instead, the tradition related to stone axe was that each individual male tomb was buried with one stone axe and it was very likely associated with his social power. In tombs dated before Fanshan tomb no. M20, stone axes were symbols of social status; the situation changed radically at M20 in which not only did the number of axes matter but also now the shape and raw material of these axes exhibited clear social differentiations.

Figure 3.14b Stone *yue* axes from Fanshan tomb no. M20

Figure 3.14c Stone *yue* axes from Wenjiashan tomb no. M1

Arc-bladed axe with a large hole ranked at the top (Figure 3.14b). Welded tuff was the favourite material and this might have something to do with its mottled texture. We note that all the stone products from Fanshan tomb no. M20 were made of welded tuff. At and around the Liangzhu site cluster there were quite a few tombs found with large numbers of stone axes: 16 pieces in Fanshan tomb no. M14, 48 pieces from Huiguanshan tomb no. M4 (Zhejiang and Yuhang 1997; Zhejiang 2014), 34 pieces from Wenjiashan tomb no. M1 (Figure 3.14c), and 132 pieces from Hengshan tomb no. M2 in Linping (Yuhang 1996). All these tombs were dated later than Fanshan tomb no. M20. Although these axes share the same form and design as the ones found in M20, very few of them were made of welded tuff. Therefore, it is assumed that it was after the 'king's' funerary assemblage represented by Fanshan tomb no. M20 that the practice of burying stone axes of the same material and shape was adopted by other Liangzhu communities around the Taihu Lake area, but most of them did not have access to such rare raw

Figure 3.15a Bi disk, Fanshan M20:186

material such as welded tuff. Only in other Liangzhu regional centres, such as Fuquanshan tomb no. M9, Sidun tomb no. M3, and a few other important elite tombs, were welded tuff axes with large drilled holes present, together with other carved jade objects, used as symbols of social status.

Fanshan tomb no. M20 also initiated the popular practice of burying jade *bi* disks. Before this tomb, jade *bi* disks were not used as a marker of social identity in Liangzhu Culture. For instance, there was not a single *bi* disk present at the Yaoshan cemetery. The design and production of typical, standard Liangzhu *bi* disk started with Fanshan tomb no. M20, in which 43 pieces of *bi* disk were recovered. Two of the *bi* disks were placed outside the burial coffin while the other 41 were deposited around the lower body, together with the previously mentioned stone axes. There was a certain pattern in the deposition of the jade *bi* disks. Those placed nearest the head were of the best quality, with a colour close to the so-called pumpkin yellow (*nanguahuang*) (Figure 3.15a). Similar material was also used in the production of the *yue* axes and *cong* tubes in the Fanshan tomb no. M12. Another 12 of the 43 pieces were made of a yellowish-brown nephrite and placed on top of the other *yue* axes, suggesting that they were used in the end of a prolonged funeral ceremony involving the use of jade. The lowest kind of *bi* disks was made from greyish-blue- and greyish-brown-coloured nephrite and placed randomly at the bottom (Figure 3.15b). These disks were very poorly made, with some not even polished (see Figure 3.2).

Dated later than M20, Fanshan tombs no. M14 and no. M23 were each buried with 26 and 54 pieces of jade *bi* disks, respectively. These disks were generally bigger in size and most made of low-quality nephrite with grey speckles. Despite this, each tomb was also buried with several elaborately produced jade *bi* disks made of yellowish-brown color and placed, as usual, separately near the head or on the upper body of the occupant.

Figure 3.15b Poor quality *bi* disks, Fanshan M20

If we plot diameters of jade *bi* disks and holes from Fanshan tombs onto a scatterplot diagram and compare them with those from large burials at Sidun (Nanjing 1984) and Qiuchengdun (Nanjing et al. 2010), we see a clearly defined pattern of the cross-cemetery differences and connections (Figure 3.15c). On this diagram, the sizes of *bi* disks from Fanshan tombs no. M16 and M12 group separately from the others. The drilled holes of those from M16 are generally large, while the sizes of the ones from M12 are generally small. Jade *bi* disks from the Qiuchengdun and Sidun sites were generally bigger than the ones excavated from Fanshan. It is very likely that they were produced by two different social groups. On the whole, there is a positive correlation between the size of the Fanshan *bi* disks and the diameter of the holes in the middle while in the case of Qiuchengdun and Sidun disks, the size of the disks was distributed relatively randomly but the diameter of the middle holes remained largely the same. There is no full dataset published on the Fuquanshan *bi* disks to date. However, based on currently available information we know that there is a separation; some of the Fuquanshan *bi* disks (Shanghai 2000) appear similar to the Fanshan ones while others have the same design as the Qiuchengdun-Sidun group, which shows that the ones at Fuquanshan might have come from diverse resources.

Cong tubes, *bi* disks, and *yue* axes were the three most prestigious objects found in the Liangzhu elite tombs. These objects have different usage and social meanings within this elite circle. *Cong* tubes held the central position in the Liangzhu belief system; *yue* axes can be divided into two subgroups, stone and jade, and both groups were indicators of male power. In comparison, the

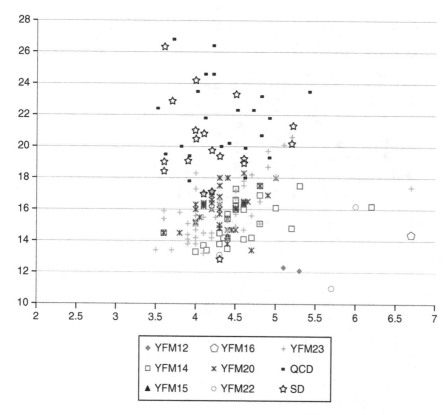

Figure 3.15c Comparison of diameters of *bi* disks (Y) and holes (X)

appearance of *bi* disk in Liangzhu Culture was relatively late and they were normally made from different raw material with other carved objects. As many scholars have assumed, the *bi* disk was probably an expression of one's wealth.

Fanshan tomb no. M20 was buried with two other important kinds of jade artefacts: Handle end ornaments and beads. M20 has the most jade end ornaments currently known, with 29 pieces in total. Some of these end ornaments were recovered in pairs when excavated. In most cases, archaeologists do not know how they were arranged and used in Liangzhu society. However, although many of the organic material on which these end ornaments were attached had long decayed, it is still possible to speculate about how these variously sized sceptre-like objects were used and to which part these end ornaments were attached. These sceptre-like objects were not part of the clothing of the deceased but were used during the funerary rites before being placed inside the coffin.

More than 300 jade beads were excavated from Fanshan tomb no. M20. As most of them were randomly scattered within the burial, it is only partly possible to reconstruct their original arrangement. One of them was made

Figure 3.16a Details of beads set with *cong*-styled tube beads (106 pieces), Fanshan tomb no. M20

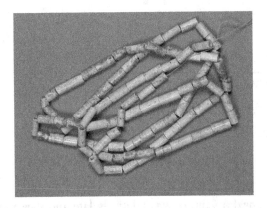

Figure 3.16b Beads set (88 pieces), Fanshan no. M20

of 106 small pieces of jade, which included four small *cong*-style tube beads (Figure 3.16a) and the second largest set had 88 beads (Figure 3.16b). Large numbers of bead ornaments, in various arrangements, were typical of elite burials at Fanshan and Yaoshan. According to the whole amount of beads set from each tomb, it is assumed that they were too many for an individual's everyday wear but only used during death rituals. Although small jade beads were also present in ordinary Liangzhu tombs, their appearance in large numbers in high-class burials, undoubtedly, suggests that the tomb occupant had strong control over precious raw materials.

Fanshan tomb no. M20 on the whole marks a changing point in the Fanshan cemetery as well as in Liangzhu society. From this tomb onwards, many new material expressions became associated with one's social position, especially large numbers of *yue* axes and *bi* disks. However, Fanshan tomb no. M20 inherited the burial pattern from the preceding Yaoshan cemetery and early phase tombs at Fanshan. Last but not least, as discussed repeatedly in the preceding text, Fanshan tomb no. M20 exhibited a wide range of connections with Fanshan tomb no. M12, the Yaoshan cemetery, and other Liangzhu high-class tombs located further away.

Within the Liangzhu Culture area, starting from the Yaoshan cemetery, a set of fixed regulations for funerary and burial rituals was gradually taking shape, including gender differences in burial patterns. During this process, the number of *bi* disks and *yue* axes became another parameter deciding one's social status within the Liangzhu system, after *cong* tubes. The close relationship between the communities or groups buried at Yaoshan and Fanshan and their general networks with other Liangzhu elite groups had all been materialised and consolidated by a rather fixed set of burial patterns and ritual practices in which jade played a central role.

3.2 Signs of power: The structure and order in the Liangzhu City

Projecting power can be achieved through multiple means. By looking at the Liangzhu locations and the excavated materials from these places, we can understand, from various perspectives, how that social power was distributed and how social hierarchical systems operated. At the same time, material remains that were associated with jade production may also be useful in understanding Liangzhu society.

3.2.1 Power and space

For a long time, the Liangzhu site cluster has remained at the centre of discussions on Liangzhu social power. However, the Liangzhu site cluster as a spatial concept has been changing with the accumulation of new archaeological material from both regional surveys and excavations. Originally estimated at 33.8 square kilometers in 1994, expanding to 50 square kilometers based upon the heritage conservation planning in 2005, and expanded again based upon the recent discovery of Liangzhu City (Zhejiang 2005c) and hydraulic system, our understanding of the whole Liangzhu site cluster structure is continuously being revised with new research (see Chapter 2 for detail).

First, it is important to acknowledge the changes that have taken place in our understanding of the chronology of Liangzhu. Through decades' archaeological fieldwork, the structure of the Liangzhu site cluster has now become clearer and we can largely divide the development of the site cluster into three phases. In the early phase, Yaoshan remained the elite burial ground and the Wujiabu location in the west and the Miaoqian location marked the eastern border, which has also yielded archaeological remains of the early Liangzhu period. Recently, archaeologists found the Guanjingtou site

(Zhao, Y. 2013), to the south of the previous known Liangzhu site cluster, which also yielded many jade items with beautiful carvings, comparable to those discovered at Yaoshan. To the east of the complex is the Yujiashan site in Linping area (see Figure 3.38 right third from top), where archaeologists found a series of burials dated from the early to late Liangzhu period. With all these new discoveries, we have had a better understanding of the early Liangzhu period archaeological remains and their spatial distribution as well as how the Liangzhu site cluster came into being in the first place.

For the middle Liangzhu period, the most important discovery was the hydraulic system. Starting from the Penggong location, archaeologists identified several earthen ridges which constituted the main structure of the Liangzhu dams. Dating results from various locations indicate that the construction of the hydraulic system was completed during the middle Liangzhu period (*c.* 3100–2900 BC). Therefore, for the Liangzhu people, who lived around that time, the functional area of the hydraulic system would have been very clear. In this context, Liangzhu period settlements along the southern edge of the Daxiongshan Mountains and in Linping area were clearly not located within the scope of the 'center of power' of Liangzhu society.

Although the date of the initial construction of Liangzhu earthen walls is unknown, currently available 14C dates and archaeological remains both suggest that the late Liangzhu (2600–2300 BC) period was the most prosperous period based upon the use of the main part of Liangzhu city walls, as well as the surrounding settlements such as the Bianjiashan site, and the Meirendi site. In this context, we may argue that in the late Liangzhu period, the social power seemed to be more spatially centralised. At the same time, late Liangzhu society seemed to have been very influential among its surrounding areas. For instance, in Linping area, the aforementioned Yujiashan site was continuously used and was in fact expanded into six enclosures (Lou et al. 2012; see Figure 2.20). Large paddy fields with well-equipped water management facilities were excavated at Maoshan (Ding et al. 2010; see Figure 2.19). At Hengshan, archaeologists found some large burials dated to the late Liangzhu period, which are comparable to those in Fanshan, showing that Hengshan was connected with the Liangzhu elite network.

In summary, when discussing the relationship between power and space, the spatial changes over time could be considered a part of this process of the establishment, consolidation, and expansion of power.

Second, we need to also consider the spatial differences exhibited between the living and the dead. There are abundant ceramics for daily use excavated from Liangzhu City, Mojiaoshan , and other ordinary settlements outside the main site cluster. These objects are as important as those jade items excavated from high-class burials, which show us a 'living space', very different from that of the dead.

One major difference between the living and the dead was how social power was expressed. Tombs of various social classes used different sets of burial objects. Such difference was expressed between but not within cemeteries. Interestingly, different social classes used the same style of daily ceramics. Whether discovered at the central platform of Mojiaoshan, located at the very centre of the site cluster, or found in the waste deposits near the earthen walls, the utilitarian ceramic items are more or less the same. For instance, ceramics unearthed from the Bianjiashan site, a location outside the walled area, are not much different from those daily-use objects found inside Liangzhu City. Many of these ceramic items from Bianjiashan also have beautiful carved symbols on their surface, the same as those found inside the city. However, a major difference can be found within the 'space for the dead'. Again, the Bianjiashan cemetery serves as a representative example of low-class tombs (see below). Thus, burials require more attention than occupational areas if one wants to examine how Liangzhu society was stratified.

The difference between the living and the dead can also be perceived in their attitudes towards certain objects. Liangzhu jade seemed to only have been used in burial contexts and is rarely found in residential locations; of course, we cannot exclude the cases in which jade was used before death, in a ritual context at residential areas. Shang and Zhou bronze hoards could be used as an inspiration for the possibility of similar process to jade after certain ceremony. Somehow, there is so far not any evidence to imply the similar situation in Liangzhu.

As for ceramics, Liangzhu people used very different sets for the living and for the dead. Pottery items found in tombs came in a very fixed group, which did not show much hierarchical differences, very different to the hierarchical system of jade as previously discussed. These pottery items were all made from a coarse clay and fired at a low temperature. In contrast, Liangzhu ceramics found at residential settlements were often beautifully produced; in particular, the drinking vessels such as cups and *hu* bottle were often decorated with carved patterns.

Another point worth mentioning is the carved patterns applied on some of the Liangzhu ceramics, which showed a close relationship with the decorative motifs found on jade yet in a quite different style. Some ceramic sherds were decorated with the similar animal-face and bird motifs (Figure 3.16, Figure 3.17), very important subjects in the Liangzhu Culture. These images or motifs, when they appeared on jade, were important symbols of social power or status. However, when these same motifs were applied on ceramics, it seems that they did not serve the same purpose. Through this contrast, we can see that different meanings exist when these motifs were used together with different types of materials. In other words, there is difference in the Liangzhu belief system between when it was practiced in the living and the dead domains.

The third and last point I want to mention in terms of understanding Liangzhu social power and space is that we need to place the Liangzhu site cluster within the Liangzhu Culture, a much larger region in lower Yangzte. I would like to refer to the term 'space' as not only space in terms of a territorial sense but also in terms of a belief and/or ceremonial sense.

Figure 3.17a Carved animal mask on pottery sherds from waste deposit of the west wall of the Liangzhu City

Figure 3.17b Carved pattern (bird head with snake body) on pottery sherds from Bianjiashan site

Figure 3.17c Carved pattern (bird and geometric design) on spout fragment of pottery *hu* kettle from Bianjiashan site, which is most common pattern widely spread around Liangzhu Culture area

First, the centre of Liangzhu social power might not be the same as the geographical centre. This is a unique feature of the Liangzhu site cluster. The main distributional area of the Liangzhu Culture was around the Taihu Lake region. This spatial pattern is fundamental constrained by the local environmental conditions (see Chapter 2). Most scholars working on Liangzhu have agreed that the Liangzhu site cluster, which is located on one corner of the Taihu Lake, was not randomly chosen. There were some important links between the location of the Liangzhu site cluster and the availability of local resources. This concept will be explored further in the 'Power and Resources' section.

Second, there were spatial differences in the display of social power and the practice of the Liangzhu belief system. Across the Liangzhu area, we see a consistent set of belief systems and lifeways with similar decorative subjects, similar material remains, and daily-use objects. However, we also see that the funerary regulations within the Liangzhu site cluster were very strict. These strict funerary rules were not seen outside the main Liangzhu region, even in elite burials outside the Liangzhu Site cluster. Inside the site cluster those beautifully made ceramics were used only in daily life no matter for elites or ordinary people. Outside the site cluster, such elaborate ceramics could also be found in elite burials (e.g., at the Fuquanshan cemetery). For jade, clear gender differences and the 'one tomb one *yue* axe' regulation were very rigid inside the Liangzhu site cluster, while the use of jade in other areas around Taihu Lake showed greater variation.

3.2.2 Power and hierarchy

To continue the analysis of the relationship between power and the life-death space mentioned previously, the social differentiation of the Liangzhu site cluster was mainly manifest through the space for the dead, namely, the burials. Therefore, for our discussion of the so-called power and hierarchy it is essential to analyse the pattern of differentiation among the burials.

According to the existing evidence, the Liangzhu burials can be divided into at least three classes or tiers: The high-class tombs are represented by the ones at the Yaoshan, Fanshan, and Huiguanshan cemeteries. The tombs at Huiguanshan are slightly inferior to Fanshan and Yaoshan in terms of the sheer quantity and quality of the burial objects, but from the perspective of the cemetery structure and the composition of burial objects, the differences between the Huiguanshan and the other two cemeteries would not differentiate them into two tiers. The tombs of the first tier were usually built on artificial platforms, indicating large labour investment. In addition to the fixed combination of pottery vessels, there were a good number of jade and stone items, including *cong* tubes, *yue* axes, *bi* disks, *sanchaxingqi* three-pronged objects, and groups of *zhuixingqi* awl-shaped objects.

The specific characteristics of the tombs in the second tier are not quite as clear as the first tier, but the representative tombs at Boyishan (Zhejiang

Figure 3.18 Burial goods assemblage from Bianjiashan tomb no. M6

2002a) and Wenjiashan (Zhejiang 2011) primarily show the existence of an intermediate tomb group. There are usually 30 to 40 burial objects in this kind of tomb, with some grander ones even having more than 100 pieces (e.g., Wenjiashan tomb no. M1), including the fixed set of pottery vessels, stone *yue* axe, and some jade items. Among the jade, there were often individual jade objects with very clear identification markers such as *bi* disks.

Compared with the second tier, there are more tombs belonging to the third tier, such as those at the Shangkoushan (Zhejiang 2002b), Miaoqian (Zhejiang 2005d), Bianjiashan (Zhejiang 2014) cemeteries, and so on. Representative examples of third tier types can be seen at Bianjiashan, for instance. Most tombs from Bianjiashan cemetery have less than ten burial objects (Figure 3.18). In addition to the fixed combination of pottery and stone axes, the buried jade objects only contained some beads and pendants, without any from the high-end jade production system.

Obviously, the division between these three tiers is very pronounced; and in fact, there is the possibility of further dividing them up. Compared with the contrast in the number of burial goods, the structural differences reflected by the composition of the burial objects is more characteristic within high-class Liangzhu tombs, which is not fully exhibited in the other two lower tiers. There are also differences between the tomb construction methods and the location of high-class tombs, suggesting that there were different social roles within the elite community.

In addition to the three prestigious cemeteries, Fanshan, Yaoshan, and Huiguanshan, that have been excavated, there are many places where *cong* tubes, *bi* disks, *yue* axes, *zhuixingqi* awl-shaped objects, and *sanchaxingqi* three-pronged objects were collected through chance discoveries within the Liangzhu site cluster. Compared with the ones from Fanshan tomb no. M20, these collected jade items are not in any way inferior in terms of their production quality. This also indicates that there are many more top-tier tomb groups within the Liangzhu site cluster than are currently known.

If one takes a rough calculation of the tombs excavated at the Liangzhu site cluster, the different tiers of tombs (communities) do not constitute a complex society, which are generally considered to be a pyramidal structure. The proportion of the elite at the top of Liangzhu society was probably not as small as conventionally thought for a pyramidal society. At the same time, even the third-tier burials, thought to belong to members of the lowest social class, were buried with the same assemblage of pottery as the elite tombs. There were also some jade ornaments buried in these tombs, suggesting that the lowest class also had access to some jade resources.

The social structure and differentiation in the site cluster may not represent the full picture of Liangzhu Culture or the society as a whole represented across the region. However, one can argue that it is due to the existence of a large number of wealthy and powerful elites that allowed for the formation of this particular structure of hierarchy among Liangzhu Site Cluster. Therefore, in terms of space and hierarchy, the central power of Liangzhu Culture was undoubtedly concentrated within the site cluster. From this perspective, the site cluster itself ruled over and constituted a special hierarchy within the whole of Liangzhu society.

3.2.3 Power and gender

The differences between male and female jade use can be clearly seen in the Liangzhu tombs, especially in the elite burials. This is plainly shown by the presence of power-related objects only in the males' tombs, including *yue* axes, *sanchaxingqi* three-pronged objects, and sets of *zhuixingqi* awl-shaped objects; meantime, in the Liangzhu female elite tombs there are certain types and combinations of objects that did not appear in male burials. However, we also need to be mindful that the use of the Liangzhu jade decorative system showed no difference between men and women, pointing to the fact that the connection between the power and gender in Liangzhu burial practices was rational and orderly, within a shared belief system.

Starting with the Yaoshan cemetery, the differences in the structure of burial objects from the north and south rows were strict and institutionalised. This distinction was also reflected at the Fanshan cemetery. Although there are no human skeleton remains, clear gender differentiation between burials

Figure 3.19a Huang semi-circular ornament, Fanshan tomb no. M22:8

in the southern and northern row at Yaoshan can be observed from the jade spindle whorl and weaving devices. For both large elite tombs and other common burials within the Liangzhu site cluster, the same pattern of differentiation also existed. The association between male tombs and the inclusion of stone *yue* axes was particularly prominent.

Tomb no. M22 at Fanshan is a typical burial for a female elite. In addition to the *guanzhuangshi* cockscomb-shaped objects, the composition of head decorations for men and women was very different. However, the fact that both men and women were buried with *guanzhuangshi* cockscomb-shaped objects, which is a jade comb head, indicates that there was probably no gender difference in their hairstyle.

Around the heads of Liangzhu females, there is usually a set of jade tubes, often with *huang* semi-circular ornaments. Together, they formed a set of tube-*huang* ornaments. In the female tombs at Fanshan and Yaoshan, there are also a few cases of jade tube beads found without *huang* semi-circular pendants. However, only one example of two sets of tube-*huang* ornaments is found in one female tomb, which is Yaoshan tomb no. M11. Item no. M22:8 consists of 12 jade tubes and one carved *huang* semi-circular ornament. This *huang* has a very special appearance and is indeed similar to the so-called semi-circular ornament (Figure 3.19a). Because they were excavated near the head of the occupant, it is unlikely that they were used as a necklace.

Item no. M22:20 is another example of a *huang* semi-circular ornament, found where the neck of the tomb occupant would have been, with the image of the sacred human with animal sculpted on the front but without the part of the man's wrist (Figure 3.19b), which directly corresponds to the 'sacred face' as seen on items discovered at Fanshan tomb no. M12. Fanshan tomb no. M22 is the female tomb out of the whole cemetery that has the closet connection with Fanshan M12, and this warrants further comparative

Figure 3.19b Huang semi-circular ornament, Fanshan tomb no. M22:20, compared with 3.19a in terms of the different way of viewing carved motif

Figure 3.19c Reconstruction of '*huang*'-tube pedants on head of female

studies. Another interesting fact of this piece is that the direction of the carved face is opposite to the one placed over the head (Figure 3.19a, item no. M22:8), indicating that the two objects are used and displayed from different positions and confirming the assumption that the *huang* with tube pendants placed near the head was not a necklace, but likely a 'crown' (Figure 3.19c).

Compared with the females, the male elites of Liangzhu had more objects with gender attributes, including the s*anchaxingqi* three-pronged object, sets of *zhuixingqi* awl-shaped objects, and the semi-circular ornaments typical of Liangzhu Culture (usually in a set of four pieces). The funeral assemblage of Fanshan tomb no. M20 is the representative of such a typical male elite burial inventory.

Another special kind of object, normally found on the chest and belly of the female occupants, in these types of burials is the rounded-shaped plaques. This type of plaque usually came in sets. They were found placed along the

Figure 3.19d Unearthed arrangement of *huang* and round-shaped plaque pedants on the neck and chest of female

chest of the decreased. In some cases, the plaque is found together with *huang* semi-circular ornament, so they are sometimes considered to be the earliest prototype of the later jade pendants set (Figure 3.19d). In the case of Fanshan M22, all six rounded-shaped plaques are decorated with dragon-heads pattern (Figure 3.19e), five of which with dragons face to face and only one with dragon heads towards the same direction (Figure 3.19f). It is unclear whether there is any special meaning of these orientations of dragon heads.

The large numbers of jade beads found in Fanshan and Yaoshan burials is truly astonishing. Most of them are composed of hundreds of small, single tubes. In some cases, these are taken as whole sets of string ornaments, and most of them are found in elite male tombs. For now, it is hard to confirm whether the tube strings are the special ornaments for men. However, apart from *huang* semi-circular ornaments with tube strings, it is very rare for female elites to have large sets of beads as burial objects.

The gender difference in women's use of *huang* semi-circular ornaments and men's burial with *yue* axes had become clear from the late Songze Culture

Figure 3.19e Rounded-shaped plaques with dragon-heads pattern from Fanshan tomb no. M22

Figure 3.19f Detail of dragon-heads pattern towards the same direction

(3500–3300 BC). The lower Yangtze River had the tradition of using *huang* semi-circular ornaments as ornaments for a long time. At first, *huang* semi-circular ornaments had no clear gender attributes. Since the Beiyinyangying and Lingjiatan cultures, where large numbers of *huang* semi-circular ornament were crafted into arc-shaped with quartz and into flat-shaped with nephrite, *huang* semi-circular ornaments became the main product of the craft industry in the Lower Yangtze River. From the late Songze period, with the *huang* semi-circular ornaments acquired more and more gender attributes, the number of findings actually decreased. During the Liangzhu period, *huang*, as a common female identity marker, was not produced in large numbers. Not only used as a single ornament, *huang* semi-circular ornaments in Liangzhu Culture had also began to be used with other string ornaments and plaque ornaments as a set pendants in female tombs. In the later Liangzhu jade system, *huang* semi-circular ornaments' position did not get strengthened but become less important, and this was actually due to the symbolic and functional transformation of jade material in Liangzhu society. The use of jade in Liangzhu Culture, especially for high-class social groups, was institutionalised and governed by rather fixed practices. In this system, decorations are certainly no longer the utmost important aspect. However, 'power', whether it was hierarchical, gender, or regional, is the essence of Liangzhu jade culture.

The differences between men and women in early cultures were also reflected in the different social division of labour. Therefore, some artefacts with characteristics of occupations have become important indicators regarding the gender of the tomb occupant. In the Liangzhu tombs, the activities related with textile production obviously directly correspond to female identity. Jade spindle whorl and weaving devices are commonly found in female tombs (e.g., Fanshan M22 yielded a jade spindle whorl). These types of jade that are related to production activities are often found in places near the feet and with other pottery, indicating that they are closer to the activities of daily life and are further removed from the funeral ritual activities.

Spindle whorls always have a special role in the research of Neolithic cultures in China. This might be because they are decorative and show geographical characteristics, thus ideal for artefact analysis. Similarly, spindle whorls appear a great deal at the Liangzhu Culture sites, and their shapes remained largely unchanged, just as those from the earlier Songze period. Although Liangzhu has yielded some stone spindle whorls, the number is quite small, and the handmade pottery spindle whorl is the dominant type. Ceramic spindle whorls sometimes have carved ornamentations, which is also a legacy from the early Songze period. A noteworthy point regarding the Liangzhu spindle whorls is that their production had nothing to do with the jade-stone industry. In Liangzhu Culture, the drilling technology was very advanced. The production of the *yue* axes and *bi* disks generated a great deal of jade cores. Similarly, a large number of stone cores were also produced when stone *yue* axes were made. Yet none of these cores were made into spindle whorls. For example, at the Wenjiashan site, 22 pieces of stone cores were excavated from settlement area, both single- and double-sided drilled (Figure 3.20) (Zhejiang 2011), but only three pieces of ceramic spindle whorls were found there and none were stone spindle whorls. The 22 stone cores bear no further processing traces (normally the diameter of spindle whorls is close to that of the holes of large stone *yue* axes). In a culture with developed jade-stone industry and advanced tubular drilling technology, almost all the shapes that can be obtained by tubular drilling are utilised in this way. Why had not the Liangzhu people seen the direct connection between the spindle whorls and the drills technology and its by-products? From the identity aspect of the spindle whorls, namely, the female

Figure 3.20 Drilled stone cores discovered at Wenjiashan

point of view, perhaps the Liangzhu economy was a highly specialised and differentiated system, in that the jade-stone industry and textile production were completely unrelated. This disassociation is not only reflected in the gender division between the labourers but also by the non-mobility of resources.

In Liangzhu society, it is not difficult to align jade and stone products with social power. Therefore, the previously mentioned two aspects of change are related to women and women's social status: The decline of female ornamental jade accompanied the development of more jade items made exclusive for male 'rituals', and the jade-stone industry also gradually alienated from everyday economic activities specialised by women (such as textile industry). Such changes are ultimately reflected in the status differences between males and females. Yaoshan was the first elite cemetery for the establishment of the male and female jade systems with the clear distinction between the northern and southern burial groups. There are six female tombs in the northern row and 774 pieces/sets of burial objects, of which 743 pieces were made of jade with no *cong* tubes found. In the seven male tombs in the southern row, if excluding tomb no. M12 (disturbed), 1,839 pieces of jade artefacts were found in these six intact male burials, including eight pieces of *cong* tubes. The differences in gender is obviously very dramatic. When the 'king's' tomb (no. M20) at the Fanshan cemetery appeared, there are only two tombs for women and seven for men in the same burial group. For Fanshan tomb no. M22, the jade carvings can be directly linked to the same style of Fanshan M12, therefore leading to the argument that M22 might be closely associated with M12. In conclusion, Liangzhu society had developed male authoritarianism, and the gender differences were closely related to the jade use system and had penetrated into the daily production activities.

3.2.4 Power and jade resource

Power often manifests itself in the control and use of specific resources. In the Liangzhu Culture, the control of jade-stone industry is not only a manifestation of social power but also a source of social power. In the Liangzhu site cluster, there are a great deal of remains related to jade production, among which the Tangshan jade production workshop was a very important one. These archaeological materials are directly related to jade production and illustrate the special status of the Liangzhu site cluster. Jade materials were obtained, manufactured, used, and distributed directly by the occupants of the site cluster. It is this direct control of resources and mass consumption of jade that made the site cluster the central power of Liangzhu society.

Chapter 4 is specifically dedicated to Liangzhu jade technology and the high-end manufacturing industry. The readers are referred to this chapter for more detail and discussion on this specific topic. Here I only make supplementary discussion of the jade workshops and other related information.

The Tangshan site is located in the north-west of the Liangzhu site cluster. A long strip running east to west, it starts with Maoyuanling in the west and Lucun in the east, with the length of 4.3 km and the width of 20–50 m. The height is 2–7 m above the surface. However, the different sections were not built simultaneously and their functions were also different. Some were built by the so-called rammed earth technology, some were piled up with sand and gravel, and some sections even had archaeological remains related to daily life and death (burials) on top of them (Zhejiang 2005c). With the discovery of Liangzhu dams such as Ganggongling (see Chapter 2), most scholars now agree that Tangshan is an integral part of the entire hydraulic system.

During the two excavation seasons at the Jincun location of Tangshan, a large number of archaeological remains related to the production of jade were unearthed, indicating that Tangshan, as well as being part of the hydraulic system, was also used as a jade production workshop probably in late Liangzhu period. For example, it was revealed in 2000 that the construction of this section was a continuous process of piling up earth, and a large amount of large stones were used to build the slope in the south. From this year's excavation, there were more than 460 pieces of jade and stone pieces. More than 400 of them are stone tools for jade production, which can be divided into three categories. First are the grindstones of different sizes and shapes. The big ones have grinding cracks across the surface, caused by grinding bar objects (such as cone or stone arrowheads) (Figure 3.21b). The intermediate ones are mostly semi-circle or cylindrical, and might have

Figure 3.21a Raw jade material and by-products excavated from Jincun jade work-shop site of Tangshan

Figure 3.21b Sandstone grounding tools excavated from Jincun jade workshop site of Tangshan

Figure 3.21c Flint carving tool excavated from Jincun jade workshop site of Tangshan

been used to polish the inner hole of jade with rectangular sides or other utensils (Figure 3.21b). The smallest ones are irregular circles, and the edge has irregular rectangular grinding surface (Figure 3.21b). The second type are remade from stone arrowheads. The shape is irregular, and most still show the original stone arrowheads form. The vast majority of them were made of tuff so that their grinding interacting surface is very smooth. It is speculated that they are polishing tools for more elaborate processing. The third kind are black quartzose stones (flint), with a few cores and a certain number of flakes. Such pointed stone chips are generally regarded as carving tools because of its hardness and form (Figure 3.21c). Jade raw materials account for the majority of the jade finds. They have different sizes, and all have cut marks, with more blade cutting marks than string cutting marks. There are also individual tubular drilled cores and broken products (Figure 3.21a), including one damaged multisectioned *cong* tube. Although archaeologists were uncertain of finding the activity surface of this workshop site, the distribution and horizontal layered accumulation of these archaeological remains make it very likely that this workshop was in use for a long time.

It is hard to determine the chronology based upon the styles and typology of the archaeological finds from this workshop, even though the unearthed tall *cong* tube remnants suggest that their date could not be too early. In addition, judging from the type and function of stone tools unearth at the site, it could be a single-step processing workshop; most of the tools are related to the grinding process. The initial cut for large pieces of jade may not have been made here. The type of the broken or unfinished jade pieces and the abrasive marks on the grinding stones indicate that different workshops might also be specialised in producing different types of jade objects. At Tangshan, there are more broken or unfinished jade pieces and tools related to the making of the *zhuixingqi* awl-shaped objects. To sum up, although the

Figure 3.22a Unfinished *cong* tubes collected from Wujiabu site

Figure 3.22b Broken piece of *cong* tube with unpolished tubular drilling marks, collected from Pingyao

site is currently one of the only excavated jade production workshop at the Liangzhu site cluster, it is not certain that its products had a direct relationship with those 'privileged goods' (*zhongqi*) found in the highest-class tombs at the Yaoshan, Fanshan, and Huiguanshan cemeteries.

However, there are other archaeological remains related to jade production found at many other sites at the Liangzhu site cluster, showing that jade production activities were quite common. Especially in the Pingyao area, unfinished plain *cong* tubes (Figure 3.22a), broken *cong* tubes (Figure 3.22b),

Figure 3.22c Tubular drill core of a tall *cong* tube, collected from the Wujiabu site

and tubular drilled cores (Figure 3.22c) were collected. These objects belong to semi-finished products, defective products, and by-products in the chain of *cong* tube production, suggesting that high-end jade production activities existed at least in the Pingyao area.

The high-class burials also have many 'shoddy' jade products, from small jade tubes even to the *cong* tubes. These 'quickly made' jade products suggest the universality of jade production activity within the Liangzhu elite circle. A good example to further look at this phenomenon is an unfinished product from Fanshan tomb no. M23. This *cong* tube is obviously an unfinished or 'semi-finished product'. The eyes of the scared human have not yet been carved, and the inner wall of the drilled hole has not been polished. It is placed with many jade objects near the feet alongside pottery containers. Obviously, this semi-finished jade could not have been used as a real *cong* tube because it was not finished. Rather, it was more likely used as a daily-used pottery or 'a large collection of *bi* disks' that are concentrated at the foot of the tomb occupant. It also suggests that tomb owner could get access to the jade products while it was in the process of being produced. Thus, high-class communities were not only consumers of jade products but also closely involved with jade production. In other words, high-end jade and stone production and consumption activities within the Liangzhu site cluster appear to have been an integrated and intertwined entity.

In contrast, the jade items and stone *yue* axes obtained by members at ordinary settlements were probably not produced by the members of these communities. At the Wenjiasha site, 22 pieces of stone drilled core were unearthed (Figure 3.20). A provenance study revealed that there are only three pieces of stone drilled core that are similar to the material used to produce the three highest-quality stone *yue* axes found in tomb no. M1. The raw material resources for the other nearly 20 drilled cores are different from the majority of the stone axes found in tomb no. M1 (Zhejiang 2011). This suggests that the majority of stone *yue* axes, especially the ones in Wenjiashan tomb no. M1, as 'imitations' of the Fanshan tomb no. M20 burial ones, are likely to be products of special production activities or from special sources, rather than locally produced.

To sum up, the relationship between power and resources at the Liangzhu site cluster is complicated and multifaceted.

1. Liangzhu site cluster presents a society of both self-production and self-consumption. Compared with Liangzhu high-class tombs, especially the Fuquanshan cemetery with obvious diversified sources of burial goods, the jade products in the entire site cluster show a great consistency in terms of material, variety, decorations, and so on. In the Fanshan and Yaoshan tombs, many cases of the jade network can be drawn between these two cemeteries. On the one hand, it shows that the relationship between the Liangzhu elites was very close and the social network was inextricably connected. On the other hand, it also reflects the concentration of jade production activities at the Liangzhu site cluster. Regarding the relationship between the jade objects found at the Liangzhu site cluster and rich burials outside the site cluster, one could find physical evidence for their direct contact. However, most of these similar jade objects have point-to-point and outwards spread relations, and there are almost no jade and stone items in the Liangzhu site cluster that demonstrate an influence from outside. The ability of Liangzhu site cluster to produce the jade and stone products is important, but more important is that the use of the jade and stone products was a very closed system, characterised by its unique pattern of self-consumption. The attitude of using their own products only is a kind of certification for the jade resources' 'legitimacy', and it also sets apart the Liangzhu site cluster from large, rich tombs at other regional centres and allowed it to become a unique centre.

2. Judging by the category of the jade items discovered at Tangshan and other locations, there were labour divisions at these jade production workshops at the Liangzhu site cluster. The division of labour can be understood from two levels: On the one hand, there are labour divisions for the production of different types of jade objects. The Tangshan workshop was mainly used to produce zhuixingqi awl-shaped objects or drills. There is no evidence for the production of large-sized items such cong tubes and yue axes. However, there are clues for cong tube production in the Pingyao area. On the other hand, the labour division is based on different stages of the production process, from the transportation of raw materials to cutting, fine processing, carving, and polishing; these steps do not seem to

be generally implemented at one place in a streamline-like style. For example, at the Tangshan workshop, a large number of grinding tools and by-products of the cutting process show the existence of the labour division for the preceding jade categories and production stages.

3. The use and distribution of jade and stone resources at the Liangzhu site cluster has many characteristics that are worthy of discussion. Generally saying in the whole Liangzhu Culture area, it is believed that the Liangzhu elites had their own circulation network for jade products; the low class could obtain certain jade resources and would also have had their own jade and stone products. In other words, Liangzhu society had a hierarchical system for the use and distribution of jade and stone products. Different cemeteries were divided into different classes based on their different roles in the production and consumption of jade and stone products. Within each cemetery there would be further divisions based on their wealth status.

    However, in the site cluster, the situation might be even more complicated. First of all, as mentioned previously, at Wenjiashan, it is likely that the burial stone *yue* axes were a special resource obtained outside rather than produced at this settlement. This might be true at many ordinary cemeteries. Therefore, this does not relate to a hierarchical system for production and consumption, nor is it a pyramid structure that relies on 'exploitation' to acquire resources. Rather, there is a layered and divergent system of jade and stone use in which core resources were distributed.

    Tangshan is another example. There are two explanations for the unfinished, semi-finished, and by-products discovered here and at a few other locations. One is that these locations were the production centres and the jade products were waste directly related to the production process. The other is that the defective or broken pieces were produced or used at other locations that re-entered the circulation network and became the raw material for producing other types of objects. Considering that most of the broken cong tubes discovered at these workshops are broken parts of finished products, the latter case is more likely. In addition, Tangshan site has quite a few examples in which huang semi-circular ornaments and bi disks were cut and remade for other objects. This indicates that after the initial consumption of high-quality jade products within the entire site cluster, there is a recollect and redistribution mechanism to continue to utilise these jade resources.

If we believe that Liangzhu social power is derived from specific beliefs and its material carriers, then the production activities of the jade industry, closely related to the belief system, are the economic base for understanding the superstructure of Liangzhu society.

3.3 Spirit of jade: Unified belief system in Liangzhu society

Liangzhu jade decoration reflects the collective recognition of the spiritual realm of the whole of Liangzhu society. The distributional area for the presence of the sacred human with animal-face pattern corresponds with that of the Liangzhu Culture sites. This spatial relationship can be seen throughout the whole Liangzhu period. This motif is not only seen on the jade but is also applied on other types of materials such as ivory, lacquer, and pottery. The combination of the sacred human and the animal face varies, but those motifs are always the core image in a piece of Liangzhu jade work that never changed.

What does this image represent? It is hard to go back to the prehistoric minds by looking at historic documents. In the book of *Explaining Graphs and Analysing Characters*, a second-century Han period Chinese dictionary, *wu* and *xi* (special agents, female *wu* and male *xi*, who could establish the connection between supernatural being and human society)[1] interpreted it as 'serves gods with the use of jade'. Perhaps the Liangzhu elites and the special community of jade users and producers are the earliest *wu* and *xi* who used jade for worshipping the gods.

3.3.1 Carving the meaning: Liangzhu's motif

The study of carved jade pattern is one of the main keys to understanding the Liangzhu spiritual world. In this section, three aspects will be explored in detail, respectively; they are the presentations of the animal face on different type of objects; the shared origin for images of dragon, human, and animals; and the diverse combinations of human, animal (monster), and bird.

First of all, the animal-face pattern as an independent decorative system emerged much earlier that the '*shenhui* insignia', so it is important to clarify that the image of 'god rides on monster' is not the same as the object of worship for the Liangzhu belief system, nor can we limit our study of the Liangzhu belief system to reading the '*Shenhui* insignia' only.

From several jade items with hollow carving and carved patterns dated to the earliest stage of the Liangzhu period, we can observe two sources for the origins of the animal-face pattern. Let us look first at some jade plaques with hollow carving. Not many Liangzhu jade objects with hollow carving have been found to date, so we can list them one by one for comparison (Figure 3.23). According to Zhao Ye (2014a), Liangzhu jade objects with hollow carvings include the *huang* semi-circular ornaments from Fanshan tomb no. M16 and Yaoshan tomb no. M11, and two old collections from *Lantian Shanfang* (private collection) and Taipei Palace Museum, respectively. In addition, there is one example from Yaoshan tomb no. M7 as a semi-circular plague and another animal face–shaped hollow carving piece from tomb no. M21 at the Guanjingtou site, located in the southern foothills of Daxiong Mountains. In the light of the crafting, the Guanjingtou one with

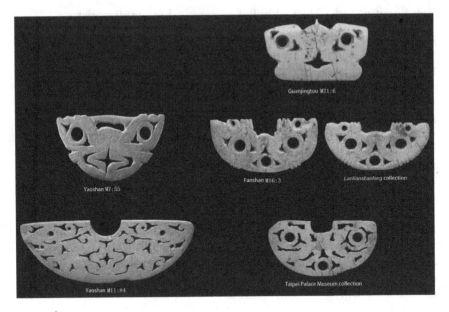

Figure 3.23 Comparison of jade objects with hollow carving

the animal face may be dated earliest among all these objects. The one from Yaoshan M7 and the pair of Fanshan M16 and its identical example from *Lantian Shanfang* can be considered as a group. The ones from Yaoshan M11 and Taipei Palace Museum, by using small solid drilled holes to position string cutting lines and design, can be classified into the third group. The difference between groups also represents technological changes over time. Two main categories of decorations can be seen among these pieces. One kind of decoration takes curved triangle patterns as its main component (Figure 3.23 right column), and the eyes and mouth (nose) of the monster are equally shown with the large round hole and triangular pattern. The other type of decorative patterns shows the eyes with large round hole and curved triangle but using cross and two 'brackets' to express the mouth of the monster (Figure 3.23 left column). The origins of the cross patterns and round hole with curved triangles can be traced back to the decorative elements commonly found on Songze-period pottery (Figure 3.24). With regard to the relationship between the Liangzhu and Songze cultures, Zhao has made an excellent summary in Chapter 5. Thus, I will skip over this topic. Based on a brief comparison of these objects, we can see that the basic components of the animal eyes and the basic expression of the whole animal face are derived from the decorative patterns on local utilitarian objects.

Another set of early jade objects with carved patterns demonstrates the homology of the animal-face and dragon-head patterns. Based on the carved

Figure 3.24 Crossed patterns and round-hole curved triangles on Songze pottery.
Top: detail of a ceramic lid from Zhaolingshan tomb no. M37 (M37:6);
bottom: detail of a ceramic handle from Zhaolingshan tomb no. M91
(M91:2)

style of the *cong* cube with accepted earliest animal face from Zhaolingshan
tomb no. M4 and the *huang*-shaped object (probably remade from a *cong*
tube) from Guanjingtou tomb no. M92, the original image of the animal
face is carved with flowing single lines. On the animal face at this stage,
similar elements and structures from the dragon-head pattern can be clearly
seen. Comparing a few examples of single line-style animal face (Figure 3.25
left) with a jade circular plaque with typical dragon head pattern from
Yaoshan tomb no. M2 (Figure 3.25 right), their facial features are almost
identical: (1) The eyes and the outer corners of the eyes are sharp triangular
spikes; (2) the line that spirals out of the eyes is finished with a sharp triangular spike; and (3) although there is no tusk in the dragon's mouth,
both animal and dragon's mouth corners continue to show the same spike.
The combination of this spike and the eyes has been noted by numerous
scholars earlier who thought it was derived from the round hole and curved
triangular pattern used in the Songze period. The special elements in the

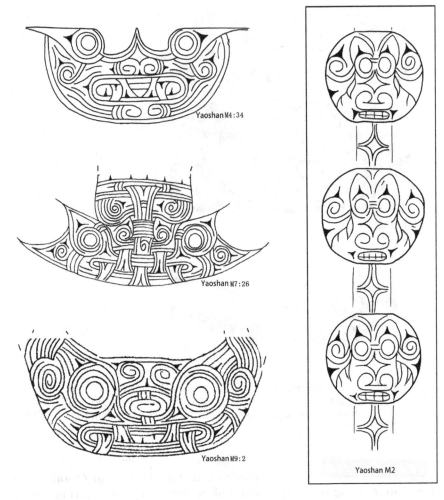

Figure 3.25 Comparison of single-line carving pattern (left) and dragon patterns (right)

dragon head pattern are the cross pattern in the middle. As discussed, it is also derived from the Songze Culture, commonly used as a basic decoration to design hollow carving patterns. In addition, for the 'garlic-shaped nose' shown by the dragon head pattern, we could immediately connect this to the similar designs seen on Liangzhu sacred human's and animal's faces.

Based upon the discussion of these two groups of objects with different decorations, it is clear that the origin of some of the basic elements of the Liangzhu animal-face pattern come directly from the Songze-period pottery decoration, including the hollow carving patterns, while the characteristics

Figure 3.26 Filled carving style of animal-face patterns carved on *zhuxingqi* cylindrical objects from Yaoshan tombs no. M9 (left) and no. M11 (right)

of the animal face pattern share an essential connection with the dragon head pattern.

The animal face carving filling the entire surface or frame of the object is basically seen on almost all important types of jade objects in the early Yaoshan tombs, including on the *cong* tubes, the *sanchaxingqi* three-pronged objects, the *guanzhuangshi* cockscomb-shaped objects, and the *zhuixingqi* awl-shaped objects. Such style is represented on several pieces from Yaoshan tomb no. M9. Different from the single-line carving, a style used to depict animal face at the very beginning phase of Liangzhu, this new fully carving style developed in two main aspects in terms of how to depict the detail of animal pattern. One way is to create and focus on the detailed description of '*yuguan* feather crest'; the other is to further expand the eye lids area to filled with detail design (Figure 3.26).

Another object that is worth mentioning is from the British Museum's Hotung collection (Figure 3.27 top). This *sanchaxingqi* three-pronged object is fully filled with a carving pattern. Compared with the *sanchaxingqi* three-pronged object from Yaoshan M10 (item no. M10:6) (Figure 3.27 bottom),

Figure 3.27 Comparison of *sanchaxingqi* three-pronged objects from the British
Museum (Hotung collection) (top) and Yaoshan tomb no. M10 (bottom)

both the subtle carving on the canthus and the details used to fill the blank
area show this consistent style. Importantly, on the object from British
Museum, the combination of dragon head, sacred human, and animal face
was presented completely at the first time. Because the profiles of dragon
noses, eyes, and nose of animals and the whole areas of sacred human with
yuguan feather crest are all presented in the bas-relief way, it is certain that
the designer or craftsmen of this piece have the clear template in mind of
this combination.

After entering the stage when the 介-shaped animal faces became popular,
the dragon-head pattern was also becoming more geometrical. It was commonly used to decorate tubes (the so-called tubes with dragon patterns),
gradually losing its direct association with the animal face. During this stage,
objects fully covered by carved patterns became less common, while decorative motifs of a single animal face or the combination of human and monster
continued to appear on various jade objects with different expression forms.
They share some common features: (1) The 介-shaped crown elements were

used to illustrate the crest or a brief symbol of both sacred human and crest; and (2) the animal eyes are not only filled with complicated designs on the eyelids but were also engraved by multiple helical and parallel coils with strings of vertical lines (the maximum number of line strings is 3). The *guanzhuangshi* cockscomb-shaped object from Fanshan tomb no. M17 shows the most typical carving style for this stage (Figure 3.28 top). The detail parts of another two objects (Figure 3.28 middle and bottom) have similar carved patterns to the one from Fanshan M17, though their sizes are much smaller. To achieve such complicated patterns of animal face within such a minute space, it really requires a remarkable carving skill. This stage of carving technology coincided with the appearance of the series of *shenhui* insignia found in Fanshan tomb no. M12. It may be closer to historical reality to understand the appearance of the *shenhui* insignia as the climax of Liangzhu carving technology. Simply speaking, the change of the Liangzhu carved animal-face pattern can be understood as the transition of the style from fully covering the surface of the jade to more focused attention on details, or the so-called micro-carving style.

The *guanzhuangshi* cockscomb-shaped objects have several distinctive features, including having inverted trapezoidal shapes and the 介-shaped notch on the central part (often containing a long oval hole). The shape of the object itself already contains the basic elements of Liangzhu motif. Unlike the other two examples of animal-face patterns, which have a pointed top (介-shaped) to represent the crown atop the forehead of the animal face (Figure 3.28 middle and bottom), this *guanzhuangshi* cockscomb-shaped object from Fanshan M7 uses both the shape and the carved pattern to express the symbolic meanings through this delicate design.

A *zhuxingqi* cylindrical object discovered in Fanshan tomb no. M12, the same tomb unearthed *shenhui* insignia type *cong* tube and *yue* axe, provides key evidence to understand the sacred human with animal-face pattern and the *shenhui* insignia design. This *zhuxingqi* cylindrical object (Figure 3.29) has the sacred human with animal-face image (A) and the image of an animal face (B). By comparing the animal faces in these two types of images, it is clear that the technologies and details are completely consistent. Therefore, it can be surmised that the craftsmen had a clear intension to repeat the same content on one object. The only difference between the two types of animal face is the forehead between the eyes. In image type A (Figure 3.29a left), the 介-shaped element at the forehead of the animal face is represented by the crest atop the sacred human, while in image type B (Figure 3.29a left), the basic 介-shaped element is reserved in the animal-face pattern at its forehead. In the image with the staggering repetition of the same patterns, the interchangeability between the sacred human with the crest and 介-shaped forehead of the animal face is obvious; both could be replaced by each other, with shared symbolic meanings.

It is worth noting that this *zhuxingqi* cylindrical object embodies the highest craftsmanship of Liangzhu, which emphasises the subject of the

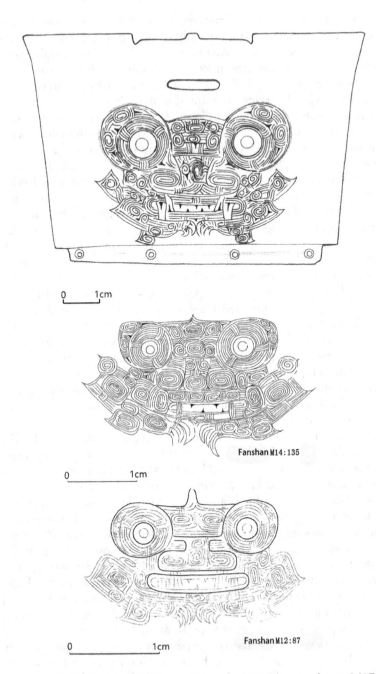

0   1cm

Fanshan M14:135

0   1cm

Fanshan M12:87

0   1cm

Figure 3.28 介-shaped animal-face patterns from Fanshan tombs no. M17 (top), no. M14 (middle), and no. M12 (bottom)

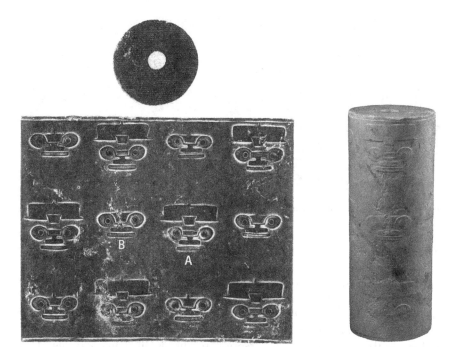

Figure 3.29a Zhuxingqi cylindrical object from Fanshan tomb no. M12, ink rubbing (left) and photo (right)

imagery by applying the bas-relief carving technology, with details of the imagery often emphasised by fine carving. This kind of imagery illustration also appears on the largest *cong* tube to date (the so-called *cong* tube king) in the same tomb (Fanshan M12) and other delicate jade objects of the Liangzhu Culture. If viewed from the perspective of ink rubbings (Figure 3.29a), the main body reflected by bas-relief carving method becomes more prominent, and the image of the animal face and the 介-shaped elements are more evidently highlighted. From this perspective, the sacred human with crest is simplified into a combination of the inverted trapezoidal face and the 介-shaped crest, which is certainly an amplification of the bridge between the eyes of the animal face in image B.

Therefore, the sacred human with animal-face pattern can be understood as a variant of the animal-face pattern. It is a more complicated expression of the Liangzhu belief system then was allowed by the development of carving technology. This new variant enriches the imagery of the animal-face pattern and provides more variables for the diversity of basic image expression. Its spiritual meaning remains unchanged, however.

Another *zhuxingqi* cylindrical object with similar layout and with the same design idea to the object from Fanshan M12 offers further insights.

Figure 3.29b Detail of cylindrical object from Hengshan tomb no. M2

The rotating pattern of the animal faces on this *zhuxingqi* cylindrical object from Hengshan tomb no. M2 is the same to those of Fanshan M12. In addition, these two objects were both placed to the right of the head of the tomb's occupant. Thus, their functions and the symbolic meanings should be the same (Figure 3.29b). The animal face of the Hengshan *zhuxingqi* cylindrical object is expressed primarily by the bas-relief carving technology without paying much attention to the detailed line carving. This unexpectedly clearly illustrated the basic composition of the animal face. Oval eyes with eyelids, garlic-shaped nose (also visible on the dragon-head pattern), long strip–shaped mouth, and, most importantly, 介-shaped elements on the forehead that are at the core to understanding the Liangzhu 'shenhui insignia'. Interestingly, this abstract version can be found directly in some examples from the same tomb, Fanshan M12, such as the semi-circular ornament (Figure 3.30 bottom). Thus, both types of abstract and more complicated 'shenhui insignia' designs coexisted in the same tomb; the conventional idea that the Liangzhu carving pattern system developed from complex to simplified needs to be revisited.

In addition to the interchangeability and mutually strengthening relationship between the sacred human and the animal face, the image of bird is also a great complement to the Liangzhu decoration and belief system. The most complete representative of the combination of man, animal, and bird is the object from Yaoshan tomb no. M2 (Yaoshan M2:1) (Figure 3.11 and 3.31). The vertical layout of the sacred human on animal-face pattern, combined with the birds on the two sides, is also seen on the *cong* tubes found in the highest-rank tombs (Figures 3.10 and 3.12).

A *guanzhuangshi* cockscomb-shaped object with hollow carving from Fanshan (Fanshan M15:7) uses different techniques to draw the picture based upon the same idea (Figure 3.32). The carved patterns show the combination of the sacred human and two birds (Figure 3.32 top); from the

Figure 3.30 Comparison of the bas-relief patterns from an object from Hengshan (top) and *huang* semi-circular ornament from Fanshan tomb no. M12 (bottom)

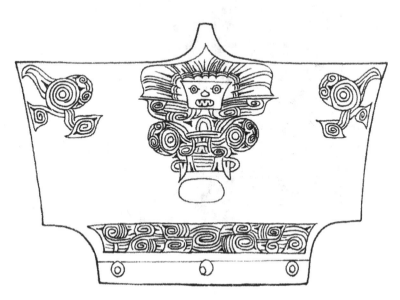

Figure 3.31 Combination of sacred human, animal, and bird on the jade comb head from Yaoshan tomb no. M2

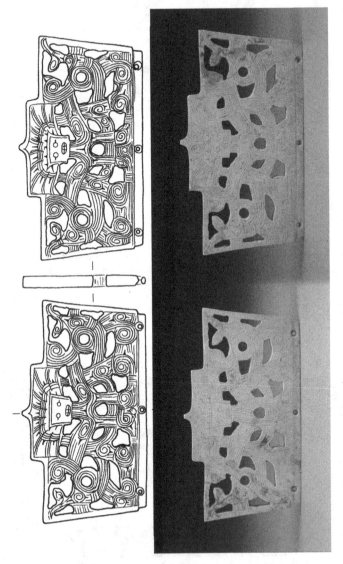

Figure 3.32 *Guanzhuangshi* cockscomb-shaped objects with hollow carving from Fanshan tomb no. M15

hollow carvings, one could recognise the basic structure of the animal face (Figure 3.32 bottom). Such a sophisticated and versatile skill for graphic transformation and expression indicates that the craftsmen had a deep understanding of the content he wanted to express.

At the end of this section, I want to briefly examine other types of materials other than jade and further demonstrate that the animal-face pattern is the true spiritual core of Liangzhu belief system. In another elite cemetery that was excavated at the Wujiachang site, next to the well-known Fuquanshan cemetery mound, not only were the high-class *cong* tubes with elaborate carved patterns recovered from M204 (see Figure 3.12), comparable to those found at Fuquanshan, a few large-sized ivory objects (Figure 3.33) covered with lacquer were also discovered in tomb no. M207. Both inside and out-side of this ivory object, ten units of the sacred human with animal-face patterns are alternately carved, similar to the design seen on the Fanshan *zhuxingqi* cylindrical object mentioned (Figure 3.29). The image is made up of the animal face and sacred human with an inverted trapezoidal face

Figure 3.33 Ivory object from Wujiachang tomb no. M207

and crest. Based on the burial pottery style, the M207 dates roughly to the late Liangzhu period. The use of the ivory object shows that the material carrier of Liangzhu ideology is not exclusively limited to jade. There are a large number of ivory remnants and lacquers found in Fanshan and Yaoshan tombs, indicating that there were indeed many perishable materials with similar images used during the Liangzhu period. The Wujiachang carved ivory objects tell us that a complete set of the sacred human with animal-face motif was still used in the practice of the late Liangzhu belief system. In other words, the specific form of the object of worship did not change in the late Liangzhu period, and the spiritual core of the Liangzhu belief system has been consistently practiced and maintained over time through the various materials.

In the Liangzhu site cluster, this particular sacred human and animal-face pattern could be carved in various types of jade objects. Sets of small tube-shaped beads, the end ornament of *yue* axes, handle of tools, belt hook, even jade spoon (Figure 3.34), all kinds of jade objects with different functions, could be used, in combination with their shapes to express the Liangzhu belief system.

Outside the Liangzhu site cluster, the sacred human with animal-face pattern also has a variety of different variations. Whether it is a complicated or abstract one, it can be only seen on *cong* tubes and *cong*-like objects (tubes, *zhuxingqi* cylindrical object, *zhuixingqi* awl-shaped objects, and so on). Different regional centres in the middle to late Liangzhu period obtained jade raw material from different sources. For example, the production at Sidun and Caoxieshan was possibly controlled by the elites, while the elite group, represented by the discovery of the Fuquanshan cemetery, acquired high-quality jade products from various sources. The jade products found at the Gaochengdun cemetery might have been directly imported from the Liangzhu site cluster made by the elite members who were buried, for example, at the Yaoshan cemetery. Regardless of the sources of the raw material or whether these jade objects were produced, the jade makers and distributors (even users) were always clear that Liangzhu motif pattern can only be carved on specific jade items (*cong* tubes and related items).

The binding relationship between pattern and shape (the jade *cong* tubes) highlights the important position of the *cong* tubes in the Liangzhu belief system and strongly suggests that *cong* remained the core material carrier in the ceremonial or funerary practice of this belief system.

3.3.2 Cong *and ritual*

Cong tubes, created by the Liangzhu people, were the product of the combination of pattern and shape. The making and use of *cong* tubes gave the Liangzhu belief system a clear 'carrier' to be adopted and practiced by the communities across the whole Lower Yangtze River delta. Eventually, the

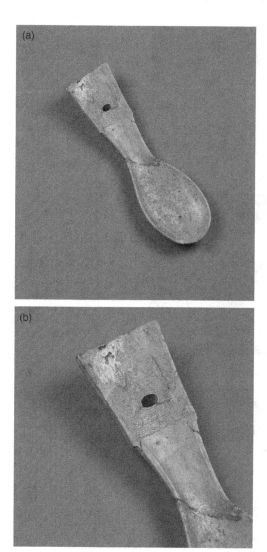

Figure 3.34 Jade spoon from Yaoshan tomb no. M12 (note detail of carving pattern shown in 3.34b)

cong tubes were separated from the spiritual soil of its origin, extending their use to the regional centres of the succeeding Longshan era and passed onto and inherited by later generations.

In the last part of this chapter, through a set of Liangzhu *cong* tubes, we revisit formation, expression, and development of this particular belief system again. Only after the meaning of the *cong* tubes is understood (the symbolism and sense of order entailed in the jade) can we start to discuss

Figure 3.35 Three largest *cong* tubes from Yaoshan (top), Fanshan (middle), and Sidun (bottom)

their profound influence on the 'ritual' of traditional Chinese society (see also Chapter 6).

The largest *cong* tube discovered in Yaoshan tomb no. M12 (Figure 3.35 top) is a masterpiece that can compete with the famous one from Fanshan (Figure 3.35 middle). Such kind of valuable jade objects make full use of the whole large piece of jade raw material and have as small holes as possible in the middle such as the ones found in Fanshan tomb no. M12 and Sidun tomb no. M3 (Figure 3.35 bottom). The tombs with this kind of jade *cong* tubes are always located at the centre of the cemeteries. Such *cong* tubes thus represent not only the social status of the tomb owner but also the social power of the whole community (family) behind them. The two *cong* tubes from Yaoshan M12 and Fanshan M12 are different in their decorative details, representative of the crafting style of their own group (cemetery). To a large extent, both items could be viewed as products of their respective family's control of jade resource. The *cong* tubes from Sidun tomb no. M3, despite being closer to the Yaoshan and Fanshan ones in its sheer volume, the raw material, and the simplified double-sectioned sacred human design, are consistent with other tall *cong* tubes in the same tomb at Sidun. Undoubtedly, these *cong* tubes are the own product of the Sidun group. Although there are other high-rank tombs in the Liangzhu Culture (e.g., Fuquanshan), only three such kind of *cong* tubes, or 'valuable items of the family', have been found so far. Associated with the diverse sources of the Fuquanshan archaeological assemblage, it can be surmised that these 'valuable items of the family' were customised products produced by the elite group who had direct control and use of jade materials and jade production.

A group of square *cong* tubes further demonstrate the combination of pattern and shape and its variations present on these objects (Figure 3.36). As discussed in the preceding text, there are diverse expressions of the sacred human with animal-face patterns on the jade items discovered at the Liangzhu site cluster. These expressions are not limited by the 'frame' (a certain area of the protruding corners on *cong*). Only in the example appearing on the section 'frame' of some *cong* tubes (represented by those from Yaoshan tombs no. M9 and M11), the design followed the principle of 'suitable for frame', namely, the deformation of a variety of basic elements is shown within a certain space.

In such cases, the 介-shaped element in the upper part of animal face is represented by crest or the crest of the sacred human (Figure 3.36a) or by the double horizontal lines which also replace the crest (Figure 3.36b). At the same time, the carving technique applied on the section 'frame' gradually transited from fully carving to highlighting the main body of the animal face (eyes, nose, and mouth). After the appearance of multisectioned *cong* tubes, horizontal lines, sacred human, and animal face became a basic combination to be carved as a fixed set of two sections of one corner (Figure 3.36c). In this combination, there is a tendency to simplify the eyes, nose, and forehead of the animal face (Figure 3.36d). Meanwhile, appearing very early, the square *cong* tubes with the sacred human on a single section and double sections had also shown simplification, but their sheer volume and use have no difference from the typical square *cong* tubes with the animal face or sacred human with the animal-face patterns. Currently, apart from re-cutting pieces, there are not real examples of *cong* tubes with three or

Figure 3.36 Square *cong* tubes from Yaoshan – 3.36a M12-2789; 3.36b M7:34; 3.36c M2:22; 3.36d M12-2787

Figure 3.37 Bracelet-like *cong* tube from Yaoshan tomb no. M9

four sections of the sacred human. Because of the lack of such intermediate examples, I always think it is difficult to say that the tall *cong* tubes with multiple sections appearing in the late Liangzhu is a development from the short square *cong* tubes with the double-sectioned sacred human pattern. The multiple-sectioned *cong* tubes in the late period have a high uniformity in their distribution in the Taihu Lake region and beyond. The nephrite material used by this kind of tall *cong* tubes is also different from that of most jade objects with carved patterns at the Liangzhu site cluster. A relatively small number (currently only three cases) of tall *cong* tubes are found at the Liangzhu site cluster, while in the high-rank tombs at the Caoxiashan, Sidun, and other cemeteries in southern Jiangsu to Zhenjiang area, they are relatively more common. From the perspective of power and belief, a new system of jade production, belief, and power differentiation must have been developed behind the production and use of these tall *cong* tubes.

The last type I want to discuss is the *zhuoshicong* bracelet-like *cong* tubes. Three typical examples of early, middle, and later Liangzhu are, respectively, from Yaoshan tomb no. M9 (Figure 3.37), Huiguanshan tomb no. M12

Figure 3.38 Comparison of early bracelet-like *cong* tubes and carved patterns – left and bottom right: Yaoshan no. M9; top right: Gaochengdun no. M13; second top right: Yaoshan no. M10; third top right: Yujiashan no. M200

(Figure 3.39a), and Hengshan tomb no. M2 (Figure 3.40b). Chronologically speaking, *zhuoshicong* bracelet-like *cong* tubes have been used continuously throughout the Liangzhu period. In term of their function, not only the *zhuoshicong* bracelet-like *cong* tubes were used as arm ornament for the disease but other arm ornaments such as the short square *cong* tubes from Xindili tomb no. M137 (item no. M137:9) (Zhejiang 2006a) and Puanqiao tomb no. M11 (item no. M11:18) (Peking University et al. 1998) had similar functions. Based on this, those round or short square *cong* tubes

Figure 3.39a Zhuoshicong bracelet-like *cong* tube from Huiguanshan tomb no. M2

Figure 3.39b Expanded view of *zhuoshicong* bracelet-like *cong* tube from Huiguanshan tomb no. M2

with appropriate size holes, placed at the upper part of the tomb owner, could be used as arm ornament as well. Therefore, neither *zhuoshicong* bracelet-like *cong* tubes nor the tube-like *cong* tubes specially for wearing can be considered as the origin for the shape and function of *cong* tubes. The function continued until the late Liangzhu period. Wearing *cong* tubes could have been part of the special spiritual practices in Liangzhu society.

From the shape and pattern, the aforementioned *zhuoshicong* bracelet-like *cong* tubes also represent the characteristics of Liangzhu social networks in different periods. An early one from Yaoshan tomb no. M9 can be used to compare with ones from Yaoshan tomb no. M10 (item no. M10:15) and Yujiashan tomb no. M200 for both shape and pattern. The same carving style can also be seen on a square jade *cong* tube from Gaochengdun tomb no. M13 in Jiangyin, north edge of the Taihu lake region. The latter is definitely an imported product from the Liangzhu site cluster (Figure 3.38 top right). This suggests that the main stream of carved jade products in this period are controlled by the Liangzhu site cluster group. This power centre displayed and maintained its belief system in jade objects through the circulation and distribution of finished products.

This *zhuoshicong* bracelet-like *cong* tube of the middle phase from Huiguanshan tomb no. M2 is the only one with five sections found so far (Figure 3.39a, 3.39b). Considering the *zhuxingqi* cylindrical objects with similar frame design always appears three sections in sets (e.g., M9:1), the

Figure 3.39c Comparison of carved eyes on the *zhuoshicong* bracelet-like *cong* tube from Huiguanshan tomb no. M2 (left) and Fanshan tomb no. M12 (right)

number of sections probably does not have special meanings. The four sections form are essentially the form to fit to the design of short square *cong* tube with its four corners. The eyes of the animal on this *zhuoshicong* bracelet-like *cong* tube from Huiguanshan M2 is finely carved, almost identical to those of a *cong* tube from Fanshan M12 (Figure 3.39c left and right). The middle phase witnessed the kingship at the Fanshan site reached its peak. It is in this period that the power of the Liangzhu site cluster is very concentrated and closed. In particular, the 'micro carving' style seen on the jade objects from Fanshan M12 had a limited distribution. Only a few elites could have had access to such jade products. A bold suggestion is that, by the middle Liangzhu period when the belief system was basically unified, the centralisation of kingship was probably the most important social development in the Liangzhu site cluster.

The Hengshan *zhuoshicong* bracelet-like *cong* tube belonged to the late Liangzhu period and can be compared with the same types of *cong* tubes from Sidun and Fuquanshan (Figure 3.40). Such *zhuoshicong* bracelet-like *cong* tubes are characterised by the tilt-up external canthus to the eyes of the animal. There is a lack of these *cong* tubes in the Liangzhu site cluster. Therefore, although the belief system and its practice were still implemented among Liangzhu elite groups in the areas around the Taihu Lake, the high-rank network has begun to decentralise. Considering the contemporary rise of the tall *cong* tubes discussed previously, the symbiosis of social power and belief was undergoing reshuffling during this period.

To sum up, a clear definition and explanation of the Liangzhu belief is far more difficult than the structural reconstruction of the belief system and its practices. Thus, the question about what the Liangzhu people believed

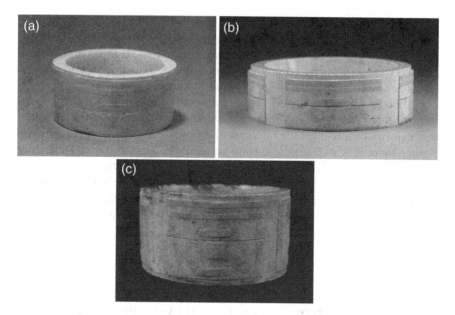

Figure 3.40 Late Liangzhu period *zhuoshicong* bracelet-like *cong* tubes – 3.40a from Fuquanshan tomb no. M9; 3.40b from Hengshan tomb no. M2; 3.40c from Sidun tomb no. M3

in cannot be answered with satisfaction. Rather, we could only reach some preliminary conclusions on the Liangzhu belief system.

1. In Liangzhu belief system, there are an abstract and unified theme and subject, which might be close to 'monotheism'. This subject is represented by the combination and the variations of the basic elements in the jade decorative system. The same imagery was expressed on other types of materials and objects as well.
2. At the Liangzhu site cluster, a strict jade burial system was established. This system was used to maintain the hierarchical order and the practice of the belief system inside the Liangzhu site cluster. At the same time, the unified 'monotheism' belief system was used to consolidate and maintain the social order within a larger social space through the pattern and shape carrier, cong tube.
3. Jade, especially its decorative motifs, shapes, and uses had reached a high degree of recognition at the Liangzhu site cluster and the wider Liangzhu area. The recognition is not only reflected in the elites' large consumption of jade but also manifested in the ordinary social groups who occasionally had access to jade resources while also strictly following the proscribed jade use system. Therefore, it was universal in Liangzhu society to practice their belief system through the use of jade.

3.4 Conclusion

The characteristics of Liangzhu jade use discussed in the preceding text provide great insights into the establishment, maintenance, and manifestation of social power in early complex societies and demonstrate the important function and meaning of a belief system to Liangzhu social order.

Although Liangzhu society was significantly differentiated, their social structure was not of a typical pyramid shape, in which the elites exploited the majority of low-class members for their own extravagant funerary rites. Rather, there was a close relationship between the Liangzhu social structure and the form of its social power, resulting from the loose and somehow mild association between the non-living resources and its belief system. This is a special feature of Liangzhu Culture that did not fully precipitate into the trajectory towards complex societies in other parts of ancient China. Giddens argues in his book *Constitution of Society* that

> power is not necessarily linked with conflict in the sense of either division that interest or active struggle, and power is not inherently oppressive ... Power is the capacity to achieve outcomes ... The existence of power presumes structures of domination whereby power that "flows smoothly" in processes of social reproduction (and is, as it were, unseen) operates
>
> (Giddens 1986, p. 257)

In a prehistoric society such as Liangzhu, this quotation is testified to some extent through jade and the practice and recognition of the belief system it supported.

It has been more than 80 years since the first discovery and research on the Liangzhu site and Liangzhu Culture in 1936. Numerous new datasets and new research methods have been produced that have broadened our vision. In this chapter and other chapters in this volume, we hope to emphasis two key concepts of the Liangzhu Culture: 'Power' and 'belief' through the study of Liangzhu society and jade, with the aim of understanding the symbiotic structure of Liangzhu social power and its belief system.

Only by understanding the distinctive characteristics of Liangzhu belief system and its related social structures can we properly place the Liangzhu Culture in the development of complex societies and early civilisations in prehistoric China. Although Liangzhu and this model of social development did not last, many of its elements have been perpetuated in various forms in later history in China.

Note

1 We hesitate here to use 'shaman', 'witch', or 'wizard' to refer to the *wu* and *xi* discussed. This is mainly to avoid unnecessary misleading interpretation. However, as the reader will find out, Sun's chapter in this volume use shaman to refer to the *wu*. We will leave this for the readers to make their own judgement and choice.

4 A controlled fine craft

Jade production techniques in the Liangzhu Culture

Fang Xiangming, translation by
Catherine Xinxin Yu

The earliest jade in the Lower Yangtze River Region can be traced back to the Kuahuqiao Culture active around 6000 BC, where three jade ornaments resembling the later period *huang* or *huang*-shaped semi-circular objects were excavated (Zhejiang and Xiaoshan 2004, p. 169). *Jue* slit rings and tubes are dominant in the Hemudu Culture (c. 5000–3500 BC) dated to a slightly later period as well as in the Majiabang Culture (c. 5000–4000 BC). Jade objects from this period were crafted from local rocks, the most prominent types from hard rocks such as quartz and carnelian and also softer rocks such as pyrophyllite and fluorite. They were flaked to form a rough shape and then polished and drilled. Artefacts related to jade production, such as raw jade, unfinished objects, and flint drills, have been excavated from a pit (H9) at the Tianluoshan site (Zhejiang et al. 2007, p. 15). A transformation in the jade culture of the Lower Yangtze area occurred around 4000 to 3300 BC. Tremolite soft jade was becoming the most common type of jade used by the Beiyinyangying-Sanxingcun Culture in the Ningzhen region, the Lingjiatan Culture around the Chaohu Lake, and the Songze Culture around the Taihu Lake region (Wen 1986). String-cutting and blade-cutting techniques assisted by abrasive sand (*jieyusha*) became more developed. The diversity of jade types increased tremendously with the appearance of *huang* semi-circular ornaments, bracelets, small round *bi* disks, and various pendant type ornaments. Jade objects also began to be closely related to settlement hierarchy, as well as the status and identity of tomb occupants. Jade culture from this period laid the foundation for the development of jade culture during the Liangzhu period.

The Liangzhu Culture was an important late Neolithic archaeological culture in the Lower Yangtze region and cast a major influence on the formation and development of Chinese civilisation. Centred around the Liangzhu City and the Liangzhu site cluster in Zhejiang Province, the Liangzhu Culture occupied an area of more than 3,600,000 ha around the Taihu Lake region and flourished for nearly a millennium. Jade embodied key social and symbolic meanings in Liangzhu Culture. Typical objects such as *cong* tubes, *bi* disks, *yue* axes, and the sacred human and animal-face motif on jade objects not only reflected the primitive religious beliefs shared by

Liangzhu society but also indicates the hierarchy of settlements containing these jade items and marks out high-status settlements that were regional centres. Mou Yongkang has discussed the mineralogical, technological, and supernatural aspects of prehistoric jade (Mou 1989a). Given that jade was a rare mineral and its production requires a vast amount of labour and socio-economic resources, a society needs to have an organised structure, a stable social order, and unanimously shared religious beliefs to ensure the successful acquisition, production, and use of jade. As one of the major fine crafts of the Liangzhu Culture (other fine crafts include ivory carving and lacquerware production, which are difficult to discuss due to the limited survival of organic matter), its systematic use reflects different levels of resource control between central, sub-central, and ordinary settlements as well as between different individual settlements. To some extent, it is also, as Chen Jie mentioned, 'economic control (production of important jade objects) and ideological control (religious beliefs that spread with jade) became the power basis of the elite in Liangzhu society' (Chen 2014, p. 290).

4.1 Jade source

Wen Guang defined 'ancient Chinese jade' as 'mainly soft nephrite', 'a tremolite-actinolite type mineral from the calcic-amphibole subgroup in the amphibole group that usually has a microscopic structure with interlocking fibres, or a nephritic structure', and 'materials similar to nephrite, in other words pseudo-jade that resembles jade, include patterned precious stones such as Xiuyan jade, xinyi jade, and some Suzhou jade that are antigorite from the serpentine subgroup' (Wen 1986, p. 42).

Other than a very small amount of turquoise, the colour of jade excavated at the Fanshan and Yaoshan sites can be divided into two main types. The first type, commonly called 'chicken-bone white (*jigubai*)', 'pumpkin yellow' (*nanguahuang*), or 'ivory white' (*xiangyabai*), is a yellowish white colour referred to as 'white' in the excavation report, *Yaoshan*. The second type is bluish-green commonly called 'duck stool green' (*yashiqing*), mainly used for *bi* disks (Zhejiang 2003, 2005a). *Cong* tube no. M12:98 from Fanshan is 'white, tenderly white, slightly pale yellow' (Zhejiang Fanshan Archaeological Team 1988, p. 10) and made of tremolite (Wen 1993, p. 133; Figures 4.5 and 4.6). Though 'dark green to green' object no. M23:167 from Fanshan is also tremolite, its iron content is obviously quite high and 'the fibre of the tremolite is notably thicker than that of the other three examples' (Wen 1995, pp. 90–97). To date, Fanshan has the highest number of jade *bi* disks, with 26 pieces from tomb no. M14, 43 from M20, and 54 from M23. Judging from their roughness and their chipped rims, Wang Mingda suggested that they symbolised wealth. Liu Bin suggested that *bi* disks did not have a fixed or irreplaceable function in the Liangzhu ritual jade system, which was dominated by objects with the sacred human and animal-face motif. 'In the Early Liangzhu period, *bi* disks were probably

Figure 4.1 Jade with various colours and in different stages of decomposition, from the Fanshan tomb no. M14 (top left, *guanzhuangshi* cockscomb-shaped object; top right, *bi* disk; lower left, *duanshi* end ornament; lower right, *bi* disk)

simply a method to show possession and use a kind of jade that was unsuitable for making any other type of jade object' (Wang 1989, p. 50; Liu 2001, p. 218).

A large number of chalky white jade items with loose structure, in the so-called *lie qin* stage[1] of jade decomposition, have also been excavated at the Fanshan site. Most of them came from burials no. M14 and M23, where *guanzhu* tubular beads, *duanshi* end ornament, *zhuxingqi* cylindrical objects, and belt hooks were found (Figure 4.1). Take *zhuxingqi* cylindrical object M23:1–3 and belt hook M14:158 as examples. Room temperature infrared spectroscopy revealed the chemical composition of the latter to be antigorite, but 'scanning electron microscopic images show that it has a loose microscopic structure. In other words, its packing density lowered significantly' (Wen 1994; Figures 4.5 and 4.6, Table 4.1).

Provenance for the raw material used for Liangzhu jade objects is not entirely clear. Meiling jade from Xiaomeiling is considered one of the possible sources (Wen and Jing 2000, p. 210; Jiang 2005; Wang 2006, p. 396). Geologists also pointed out that the Tianmu Mountains have the correct geological conditions for a jade deposit. Jade from Xiaomeiling belongs

Table 4.1 Major types of jade objects discovered in each site cluster

Types of jades \ Site cluster	Liangzhu	Linping	Tongxiang-Haining	Qingpu	Wuxian-Kunshan	Changzhou
cong, bi, yue	■	■	■	■	■	■
yue with end ornaments on handle	■			■	■	
set of semi-circular ornaments	■			■	■	
Guanzhuangshi cockscomb-shaped object	■	■	■	■	■	
Sanchaxingqi three-pronged object	■	■	■	■		
Sets of zhuixingqi awl-shaped objects	■		■			
Sets of zhuxingqi cylindrical objects	■	■	■			

Note: White indicates a small amount or complete lack of the jade type.

to the category of 'mountain jade' (*shan liao*), which cannot explain why polished concave surfaces are still visible on some Liangzhu jade objects. Raw jade samples excavated from the Jincun location at the Tangshan site in the Liangzhu site cluster show that the raw material used by the Liangzhu Culture could have been river gravel jade, similar to the so-called mountain stream or river-polished types of jade.[2] For example, raw jade T1(3):118 from Tangshan retains the natural shape of the gravel instead of being cut into shape, despite not being very rounded (Figure 4.2 top). Raw jade T12(4):19 from Tangshan almost completely retains the gravel surface and is only partially abraded to reveal the jade colour of the core (Figure 4.2 bottom).

Tongxingqi tubular objects and *cong* tubes mainly used as ring bracelets are the largest Liangzhu jade objects. Due to the limited size of raw jade, these objects often have dents and imperfections on the outside. Sometimes even the decorations cannot be carved correctly due to the imperfection of the raw jade. For example, decorations on the section surfaces of *cong* tube M7:50 from the Yaoshan site 'have different sizes due to the irregularity of the raw jade' (Zhejiang 2003, p. 81). Some scholars thought this kind of flaws were intentional, but this is actually a misunderstanding (Figure 4.3).

Figure 4.2 Two samples of raw jade from Tangshan (front and back views)

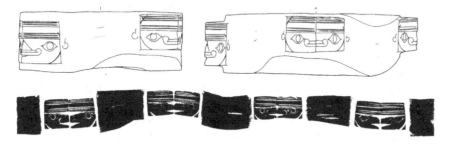

Figure 4.3 *Cong* tube no. M7:50 from the Yaoshan site and an ink rubbing of the object

Take object M12:98 from Fanshan as another example. There are indentations in the lower part of this object but decorations were still carved on the dented surface. Reaching 0.09 m tall and more than 0.17 m wide, it must have been made from the largest piece of raw jade obtainable at the time. It uses jade of the highest quality and is the most exquisitely carved among all non-flat jade objects discovered to date. The Liangzhu

Culture had a preference of using local raw material. In addition, sites such as those within the Liangzhu site cluster were probably located quite close to jade deposits. A large number of jade objects made with a similar kind of material as those found at Yaoshan were recently excavated from the Guanjingtou site in the southern part of the Liangzhu site cluster. The elite class from the Liangzhu site cluster controlled the lucrative resource of jade during the early Liangzhu period (Zhao, Y. 2013, 2014b, p. 8; Song 2014).

During the late Liangzhu period, a drastic change occurred to the types of materials used to make jade objects at the Liangzhu site cluster and nearby high-ranking settlements. In tomb no. M4 from the Houyangcun site at the Liangzhu site cluster, three *cong*-style *zhuxingqi* cylindrical objects, placed at an equal distance from one another on top of the wooden coffin lid, were so eroded and fragile that it was difficult to pick them up during the excavation. The *sanchaxingqi* three-pronged object near the head of the deceased was in a similar condition. The jade *yuntianqi* (harvesting tool) with brownish green colour could not have been more different from the ivory white jades from the Fanshan and Yaoshan sites (Wang 2009, pp. 131–132). Thirty-three *cong* tubes excavated from tomb no. M3 at the Sidun site came in all shapes and colours, including white with black spots, greenish brown, greyish white, dark bluish green, greyish green, and dark green. Object M3:22 is a tall, dark bluish green *cong* tube with 13 sections. Although it was also made with 'soft jade (nephrite)' (Nanjing Museum 1984, p. 128; Wen 1986, p. 44), it clearly came from a different source compared with the jade from the Fanshan and Yaoshan sites. The size of jade objects increased drastically. For example, *cong* tube M3:26 from the Sidun site is 36.1 cm tall and 7 cm wide. *Yue* axes from the Fuquanshan site and the Wujiachang site exceed 30 cm in height and 15 cm in width at their widest points (Shanghai Museum 2014, object M204:28 from the Wujiachang site). Nine vertically arranged *bi* jade disks excavated from the Qiuchengdun site have diameters exceeding 20 cm. M5:26 in particular has a diameter of 26.4 to 26.8 cm (Jiangsu and Wuxi 2010, p. 156).

Compared to jade excavated from central and high-ranking settlements, those from peripheral and second rank settlements are of uneven quality, which demonstrates the connection between resource control and settlement rank. During the Liangzhu period, hard stones such as quartz and carnelian were rarely used: Quartz objects almost disappeared completely, while only a very small amount of carnelian has been discovered. These stones were chipped to form a rough shape, which is completely different from the technique of using abrasive material to cut tremolite-nephrite. As for pseudo-jade from this period, there were also pyrophyllite, fluorite, and turquoise in addition to antigorite. The richest finds of pseudo-jade objects are in the Jiaxing area in northern Zhejiang Province, including shapes such as *guanzhu* tubular beads, *guanzhuangshi* cockscomb-shaped objects, and *sanchaxingqi* three-pronged objects (*cong*-style *zhuxingqi* cylindrical objects made with pyrophyllite have even been found at the Yujiashan site in Yuhang, located between the Liangzhu and Tongxiang-Haining site clusters). Because there is no pyrophyllite deposits in the boggy Jiaxing area, they must have come from mountainous areas elsewhere through the process of bartering, exchange, or gift-giving. Regardless of their origins, they are of an inferior or lower grade.

In recent years, we collaborated with the Shanghai Institute of Optics and Fine Mechanics of the Chinese Academy of Sciences, Fudan University, and the School of Archaeology and Museology of Peking University to perform trace element analysis on jade objects excavated from the Fanshan, Yaoshan, Gaochengdun, and Yujiashan sites. Our preliminary conclusions are as follows:

> It is difficult to distinguish data from the Yujiashan site. Instead, elements of the jade from Yujiashan can be grouped together with the Fanshan, Yaoshan, and Gaochengdun sites. This shows that carved jade objects in the Liangzhu Culture were made with raw materials from the same sources. Though of course, every site also has some special samples with different provenances.
>
> (Gu 2009; Gan et al. 2011)

If trace element data of jade from each regional centre or site group can be roughly established and combined with object type, shape, and decoration, then the interactions and relationships of control between these settlements should become clear.

4.2 Cutting, tubular drilling, and boring

The biggest difference between stone tools and jade products is that the former are production tools while the latter are not. As for the difference between working tremolite-nephrite and working other stones, the latter was usually flaked to form a rough shape, while the former was cut using the aid of abrasive sand. In the Majiabang-Songze periods, *jue* slit rings, tubes, and *huang* semi-circular ornaments were mainly made with quartz, which is formed into a rough shape by flaking. For example, at the jade workshop at the Fangjiazhou site, thin slabs of quartz were flaked to form a rough shape for making *jue* slit rings and *huang* semi-circular ornaments, or flaked to a cylindrical shape for making tubes, and then polished (Figure 4.4). This wasteful and imprecise method was not used to work tremolite-nephrite jade. During the Songze-Liangzhu periods, methods for acquiring a rough shape included string cutting, blade cutting, and tubular drilling.

Zhang Min is one of the earlier scholars to commence systematic investigations into jade production in the Lower Yangtze Region. In 1984, he identified straight line cutting (blade cutting), curved line cutting (string cutting), and tubular drilling as the three methods to cut jade, based on three jade production samples excavated from the Mopandun site (Figure 4.5) (Zhang, C. 2013, pp. 195–202).

Figure 4.4 Quartz in the initial stages for making *jue* slit ring, tube, and *huang* semi-
circular ornament, from the Fangjiazhou site in Tonglu

After the excavation of the Fanshan and Yaoshan sites, Mou Yongkang
wrote several articles sharing in-depth observations and analyses on cutting
methods used for Liangzhu jade objects. In addition to describing the
differences between string cutting, blade cutting, and rotary wheel (*tuoju*)
cutting, he also theorised about the three types of blade cutting (Mou
1989a, 2003).

4.2.1 Cutting

Due to the differences in the way a hard tool and a soft tool move during
the cutting process, cutting with a solid blade and cutting with a flexible
string produces different kinds of surfaces: One straighter and neater while
the other is undulated. The limitation of each cutting tool leaves distinctive
marks during the shaping and refining process (Fang 2013a).

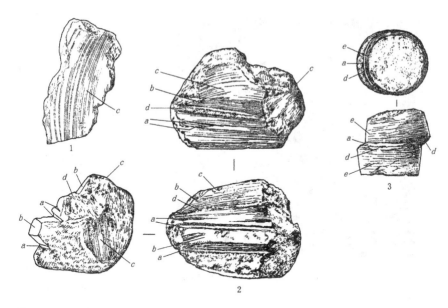

Figure 4.5 Jade-carving marks on the raw material from the Mopandun site in Dandu according to Zhang Min's observations in 1984 (1. jade cutting with a curve line mark; 2. jade cutting with a straight line mark; 3. tubular drill core; a. cutting mark; b. straight line cut; c. curve line cut; d. fracture; e. tubular drill mark)

4.2.1.1 *String cutting*

String cutting was mainly used to make jade objects with a broad flat surface, such as *yue* axes and *bi* disks. It was also used to shape the curved edges of flat jade objects, such as the semi-circular indentation of *huang* semi-circular ornaments (tubular drilling can also be used), the prongs of *sanchaxingqi* three-pronged objects, and the curved top of *guanzhuangshi* cockscomb-shaped objects. String cutting can also be used to make openwork jade, referred to as the *xiansou* technique (lit. 'sawing with string').[3] The flat Liangzhu jade *bi* disks were all string-cut, as were *yue* axes. *Yue* axe M9:14 from Yaoshan is an exception. The blade-cut mark might be the result of previous cuts. String-cut marks remain on one or both sides of many *bi* disks excavated at the Fanshan site, making these disks uneven if placed flat on one side (Figure 4.6).

The technique of using a string to cut perpendicular to the contact surface of a piece of shaped jade already existed by 6000–5000 BC, as shown by cuts on *jue* slit rings from the Xinglongwa-Xinglonggou Culture and the Early Hemudu Culture. Because the craftsman can easily control the cutting direction of the flexible string, this method was very convenient

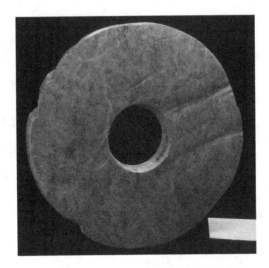

Figure 4.6 String-cut mark on *bi* disk M14:228 from Fanshan

for cutting geometrically complex shapes and openwork. The technique of first using a tubular drill or a solid drill to make a hole, and then cutting the jade with a string to make openwork, is completely different from the technique used to make openwork 'hook and cloud' ornaments in the Hongshan Culture. The latter was made by repeatedly cutting back and forth with a short blade to form grooves and holes. Thus, based upon the analysis of current evidence, the widespread use of *xiansou* (string-cut openwork) technique probably first appeared in the area between the southern slopes of the Dabieshan Mountains and the Wanjiang River region. These sites in this region that provide such evidence include the Gushan, Xuejiagang, and Lingjiatan sites (Hubei 2001b; Anhui 2004, 2006a). The contour of the jade eagle and jade human figurine found at the Lingjiatan site were likely shaped using the *xiansou* technique (Figure 4.7).

Xiansou string-cut jade objects in the Liangzhu Culture, especially openwork jade, all belonged to the early Liangzhu period. Current finds are concentrated within the Liangzhu site and around the Zhaolingshan site in southern Jiangsu Province. *Huang* semi-circular ornament M11:83 with ridge-shaped top and animal-face motif and *huang* semi-circular ornament M4:34 with ridge-shaped top from the Yaoshan site, as well as openwork *huang* semi-circular ornament with ridge-shaped top from the Guanjingtou site, look identical to the ones excavated in the area between the southern Dabieshan Mountains and the Wanjiang River region (Figure 4.8). Animal-shaped flat jade ornament M84:1 and relevant string-cut jade objects found in tomb no. M77 at the Zhaolingshan site dated between the late Songze period and early Liangzhu period, also seem to be influenced by the dissemination and interactions of jade production technologies

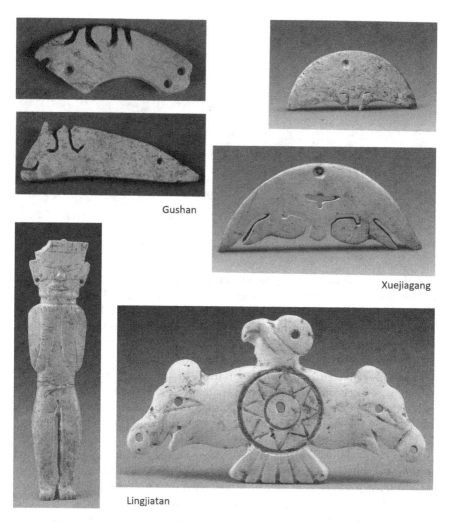

Gushan

Xuejiagang

Lingjiatan

Figure 4.7 Jade ornaments made with the *xiansou* string-cut openwork technique, from the Gushan site, the Xuejiagang site, and the Lingjiatan site

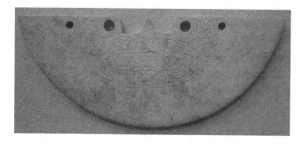

Figure 4.8 *Huang* semi-circular ornament M4:34 from the Yaoshan site

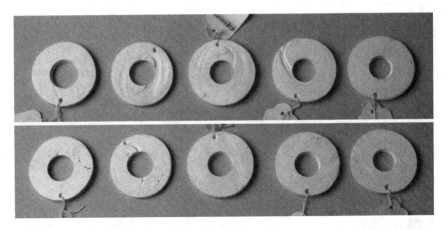

Figure 4.9 Five sets of matching flat round rings from Yaoshan tomb no. M11

mentioned previously (Nanjing Museum 2012, plate 70). This shows that connections in jade craft already existed during the early Liangzhu period between the Liangzhu site cluster, distributed around the Dongtiao River, and the Caoxieshan-Zhaolingshan site cluster, located on the banks of the Wusong River.

The decline of the *xiansou* technique after the early Liangzhu period was closely linked to the standardisation of the Liangzhu jade industry, which was becoming increasingly stagnant. The number of jade object types decreased as the Liangzhu Culture developed. For example, tomb no. M12 at the Fanshan site has 20 types of jade objects, while tomb no. M3 at the Sidun site only had *cong* tubes, *bi* disks, *yue* axes, and *guanzhu* tubular beads. However, this kind of string-cutting technique was inherited and impeccably reproduced by the Shijiahe Culture and Longshan Culture (Institute of Archaeology 1990). It was used on the rectangular openwork plaque W71:5 from the Xiaojia Wuji site, as well as on the head ornament W6:15 and W6:60 from the same site, comparable to objects M202:1 and M202:2 found at the Zhufeng site in Shandong Province (Jingzhou Museum et al. 1999, pp. 327, 331).

A set of 12 round plaques excavated in tomb no. M11 at the Yaoshan site were made by shaping the jade into a cylinder, hollowing it with a tubular drill, and then slicing it into plaques with a string tool (Figure 4.9). The vast majority of jade tubes were also cut using the same method. For someone who had mastered the technique of string cutting, the cutting direction and result can be very tightly controlled. Take the split *cong* tubes M2:14 and M2:21 from the Hengshan site as an example. Based on the height of the equally divided sections, one can reconstruct that the width of the string-cut part (as well as the stretch of space polished afterwards) was approximately 3 mm (Figure 4.10) (Yuhang Committee 1996, p. 71).

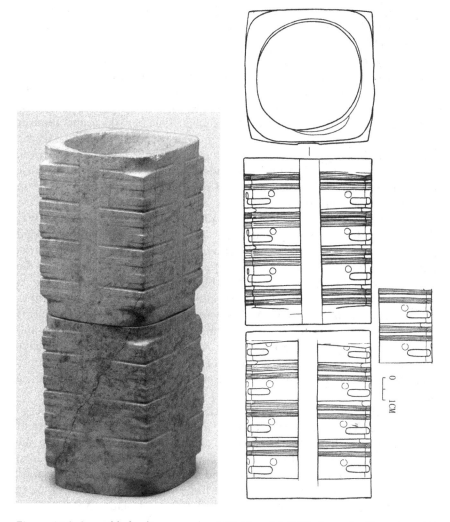

Figure 4.10 Assembled split *cong* tubes M2:14 and M2:21, from the Hengshan site in Yuyao

4.2.1.2 Blade cutting

Unlike string cutting that uses a soft flexible tool, blade cutting uses a hard tool that sometimes can also be slightly flexible. For instance, I have found that some cut marks have a curly shape which would have been created by cutting tools such as bamboo when used together with sand and water. The back-and-forth sawing motion of a blade with either a long or short edge creates products with neat and even surfaces compared to the products created when using undulating movement of the string. Given that both blade cutting and string cutting use a similar amount of time according to the results of experimental archaeology (Zhang et al. 2006a, p. 298, 2006b, pp. 315, 321), blade cutting should have an obvious advantage over string cutting, except when making the curved contour of certain jade objects and creating *xiansou* openwork. By producing flat surfaces, blade cutting can even save time and effort on polishing. Why then are there string-cut marks on many Liangzhu jade objects such as *cong* tubes, *bi* disks, and *yue* axes? String-cut marks on both sides of the exquisite jade *bi* disks found alone at Fanshan had been almost completely removed by polishing. Why was the blade-cutting technique, which was as highly developed during this period, not used instead to make Liangzhu jade objects? Clearly it was because blade cutting can only reach a limited depth, resulting in more waste in terms of time and effort.

We discovered that the depth of blade cutting during this period was very limited. It was mainly used to cut out the rough shape of elongated jade objects, such as tubes and *zhuixingqi* awl-shaped objects. Marks left on raw jade T1(3):118 from the Jincun location at Tangshan of the Liangzhu site cluster are all shallow blade-cut marks, showing that its intended use was to make *zhuixingqi* awl-shaped objects or tubes (tubes would need further cutting). A group of *zhuixingqi* awl-shaped objects M20:72-1~8 from Fanshan with almost circular cross-sections, and square *zhuixingqi* awl-shaped objects M20:67 from Fanshan, still have lengthwise blade-cut marks (Figure 4.11). The curved sides of some tubes were mistaken as the result of string cutting, but they were cut with a short blade.

Unlike the Liangzhu Culture, the preceding Lingjiatan Culture achieved much more depth with blade cutting. In addition to a small number of *yue* axes cut with blades, large stone adzes were also shaped using the blade-cutting technique. Examples include stone slabs (half-finished adzes) 98M20:10, 98M20:38, 98M20:54, and 98M20:55 found at Lingjiatan. Object 98M20:55 is 10.3 cm wide and has a wavy lengthwise mark more than 5 cm deep, left by cutting with a short-edged blade (Anhui 2006a, p. 220). This is not an isolated case because many flat jade objects from the Hongshan Culture, such as *gouyunxingqi* hook-and-cloud ornaments and jade phoenixes, were shaped with a blade that cut up to 5 cm deep. For example, grooves were cut on the backsides of jade phoenix N16M4:1 and *gouyunxingqi* hook-and-cloud ornament N16M12:1 from the Niuheliang site. They certainly were not marks left by separating the jade core from the non-precious rock but were rather grooves that facilitated drilling holes. Judging from the neat flat surfaces, these two objects were cut using a blade. Jade phoenix N16M4:1 from the Niuheliang site is about 20.43 cm long and 12.71 cm wide. Hook-and-cloud ornament N16M2:1 is 22.5 cm long and 11.4 cm wide. Hook-and-cloud ornament N2Z1M27:2 from the Niuheliang site is 9.8 cm long, 28.6 cm wide, and only 0.5 cm at

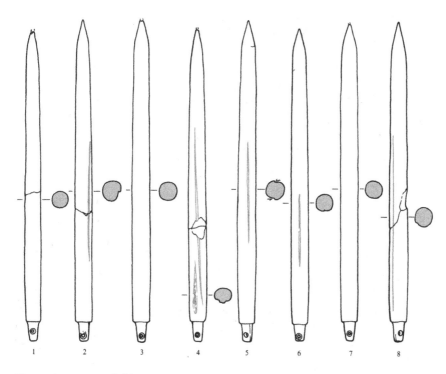

Figure 4.11 Vertical blade-cut marks on the surface of *zhuixingqi* awl-shaped ornaments from the Fanshan tomb no. M20 (1-8: M20:72-1 to M20:72-8)

the thickest point. 'It is the largest and most representative jade object excavated from the Niuheliang site and demonstrates the most skillful use of a combination of blade-cutting, groove-making, and drilling techniques' (Figure 4.12) (Liaoning 2012, Figure 71 on p. 403, Figure 85 on p. 418, Figure 65 on pp. 111 and 113).

The Longshan Culture in Shandong Province, which postdates the Liangzhu Culture, also inherited the blade-cutting technique and used it advantageously. A 'stone shovel' (Liu 1972), 48.7 cm long and 12 cm wide, was discovered by Liu Dunyuan in 1963 at the Liangchengzhen site. Although no cut mark survived, it was most likely cut with a blade. Liu Dunyuan later added an incomplete object, 'flat jade axe with holes' (Liu 1988), 30 cm long and 9.5–10.2 cm wide, that did clearly show blade cut marks. As for the large *bi* disk with a 30 cm diameter from the Qijia Culture in north-west China, and the large jade blade and *yazhang* ceremonial notched tablet from the Sanxingdui site[4] in south-west China dated to the late Shang and early Zhou period, the depth and precision achieved by blade cutting are self-evident.

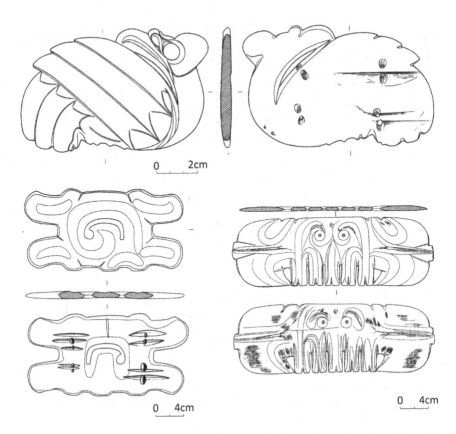

Figure 4.12 Three jade objects shaped by the blade-cutting technology, from the Niuheliang site (top, jade phoenix; lower left, hook-and-cloud ornament; lower right, hook-and-cloud ornament)

Clearly, the Liangzhu Culture did not manage to make deep cuts with blades because they were limited by the tools. The blade-cut surfaces of raw jade from the Tangshan site of the Liangzhu site cluster are neither flat nor straight, indicating that the craftsmen probably also used bamboo in addition to hard-cutting tools, the former being tough but not rigid and having a limited ability to cut deeply. However, though the depth of blade cutting was limited for the craftsmen, the length was not. Some *zhuixingqi* awl-shaped objects from the late Liangzhu period exceed 30 cm in length. For example, the *zhuixingqi* awl-shaped objects from the Dafen site in Jiaxing is 38 cm long (Lu 1991), M173:15 from the Xindili site is 33 cm long, and M9:7 from the Fuquanshan site is 32.5 cm long (Shanghai 2000, p. 85). This length is matched by that of tall *cong* tubes with multiple sections that appeared during the late Liangzhu period. It is notable that though these late Liangzhu jade objects are not comparable to those found at the

Fanshan and Yaoshan sites in terms of quality, their size far exceeds the earlier jade pieces.

While the development of blade cutting in the Liangzhu Culture was limited by tools, this technique still became very useful in other areas of jade production.

First, it was used to cut curved shapes such as the contour of *bi* disks and *huang* semi-circular ornaments. Marks showing blade cuts from two directions, preserved to different degrees, can still be detected on the outer edge of most *bi* disks excavated from the Fanshan site. The technique was also used to shape flat tenons, such as the tenon on *guanzhuangshi* cockscomb-shaped objects and the flat tenon of *mao* and *dui* end ornaments of *yue* axes.

Second, blade cutting was used to 'reduce the background (*jiandi*)' and to 'create grooves' (*dawa*).[5] Examples include raised animal-face motifs, raised nose tips on the motifs decorating *cong* tubes, bas-reliefs of the sacred human and animal-face motif, and grooves on the large animal eyes of animal-face motifs. Object M23:126 from Fanshan is an unfinished *cong* tube. The outer side of the nose has obvious cut marks left by a short-edged blade. The nose and the unfinished bowstring pattern were originally at the same height, but the raised effect of the nose was achieved by carving away the surrounding area (Figure 4.13).

The sacred human and animal-face motif on the vertical grooves of *cong* tube M12:98 from Fanshan was made in the same way. Though the vertical grooves look sunken, the highest points on the grooves match the highest points on the curved corners of the *cong* tube, the sunken effect was created by reducing the height of the two sides of the grooves. This way, the 'raised' bas-relief maintains the same height as the original round surface. The zoomed-in image clearly shows that the bas-relief was created by repeatedly cutting the background with a blade, in the same manner that seals are made (Figure 4.14).

Figure 4.13 Sunken ground of the corner motif of *cong* tube M23:126 from the Fanshan site

Figure 4.14 Traces of blade cutting used to reduce the surface for the bas-relief 'sacred man and animal-mask' motif on *cong* tube M12:98 from the Fanshan site

Figure 4.15 Sunken grooves of the large eye of the animal-mask motif on *cong* tube M12:98 from the Fanshan site

A similar technique was used to create grooves on the large animal eyes of the sacred human and animal-face motif. Shallow depressions around the multiple eye circles and crescent-shaped eyelids (or ears) of the animal eyes on *cong* tube M12:98 from Fanshan (Zhejiang 2005a, p. 58) were made with repeated and concentrated cuts from multiple directions (Figure 4.15). This is identical to the basic technique used to make grooves on hook-and-cloud ornaments in the Hongshan Culture as well as to make grooves in preparation for drilling holes on large flat surfaces of Hongshan jade objects. The fact that this technique was used to make suspension holes on many Hongshan flat rings and *bi* disks shows the close connection in jade production techniques between these two areas. The tools used for this technique were most likely short-edged blades, different from the usual blades used to cut jade into rough shapes. These techniques of removing background (*jiandi*) and creating grooves (*dawa*) shaped the grooves and ridges on Hongshan *gouyunxingqi* hook-and-cloud ornaments, while no fine incision

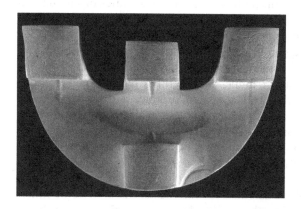

Figure 4.16 Back of the *sanchaxingqi* three-pronged object M14:135 from the Fanshan site (top), and the protrusion on *duanshi* end ornament M10:18 from the Yaoshan site (bottom) shaped by reducing the surrounding surface

lines can be seen. Carved arm ornament M458:2 from the Dadianzi site also used blade cutting to reduce background and create grooves (Institute of Archaeology 1996, pp. 171–172). Relief inscriptions on jade objects from the Shijiahe Culture also show continuity through the use of this technique.

The deep concavity of some finial ornaments and the protruding parts at the back of three-pronged objects demonstrate the most striking use of blade cutting to create relief. There are protruding parts on the upper surface of *duanshi* end ornament M10:18 from the Yaoshan site with guiding lines on the protruding parts, and clear blade-cut marks can be observed on the lower surface (Zhengjiang 2003, p. 144). Three-pronged object M14:135 from Fanshan has four protruding parts on the back side. There are vertical blade-cut marks below the top three protruding parts. The olive-shaped indentation between the protruding parts, possibly misinterpreted in the report at the time of discovery (Zhejiang 2005a, p. 95), should be the result of the abrasion of many repeated blade cuts. The string-cut mark on the right side of the lower protruding part shows that after the cutting had been completed to a certain degree, string cutting was used in small areas (Figure 4.16). Given that such extensive use of blade cutting to create relief was time-consuming and labour-intensive, objects like this were rare and were mostly found within the Liangzhu site cluster and nearby sites.

In addition, blade cutting was also used to shape jade objects with concave curves. Most *duanshi* end ornaments, some *zhuxingqi* cylindrical objects, cylindrical ring bracelets, and trumpet-shaped tubes tend to have concave contours. Given that tubular drilling would produce a drum shape with straight sides, concave curves would have to be shaped deliberately after the initial drilling (so are the concave sides of *guanzhuangshi* cockscomb-shaped objects and *yue* axes), most likely by cutting with a blade. A horizontal blade-cut mark can be seen on the slightly concave side of the cylindrical object M10:2-2 from the Yaoshan site. *Zhuxingqi* cylindrical object M2:16 from the Yaoshan site also has concave sides. 'There is a circle of rather deep but discontinuous cuts on the surface. The cuts are triangular in shape, with thin horizontal cuts across the cut surfaces.' The sides of the *duanshi* end ornament M2:15 from Yaoshan are also slightly concave. 'The surface of the object has similar cut marks as *zhuxingqi* cylindrical object M2:16' (Zhejiang 2003, p. 46). The left side of the line drawing of cylindrical object M2:16 from the Yaoshan site illustrates the horizontal cut on the cylinder's surface, showing that the cut was not the result of rotating blades. These blade-cut marks in the middle of cylindrical objects were not caused by craftsmen intending to cut the object into more segments but were the result of shaping the concavity of the surface (Figure 4.17).

Long tube M2:7 from Yaoshan has two and a half sets of dragon-head motifs, each separated by a bowstring motif, a line composed of a set of short, straight, and discontinuous streaks made by repeated blade cuts (Figure 4.18). There is good reason to believe that the bowstring lines on bowstring motif tubes, such as M16:84~86 and M16:93 from the Fanshan site, were made by blade cutting. The groove in the middle of a crystal earring 87M15:34 and the lines on bowstring motif tubes 87:15:52~63, both from the Lingjiatan site, were probably made using the same technique.[6]

4.2.2 Tubular drilling

Tubular drilling cuts out or extracts a core using a revolving tool and the aid of abrasive sand. The application of tubular drilling in the production of Liangzhu jade was quite extensive, which included drilling holes, shaping objects, making tenons and mortises, and carving the eyes of the sacred human and animal-face motif. Judging from cores and drill marks, though tubular drilling can be carried out from one or two directions, it was always achieved by single-direction drilling, performed once or twice. There was no simultaneous drilling from two opposite directions.

The potter's wheel was already used in ceramic production around 4000 BC, between the Majiabang period and early Songze period, while the fast wheel was also rapidly developing. During the Songze-Liangzhu period, the

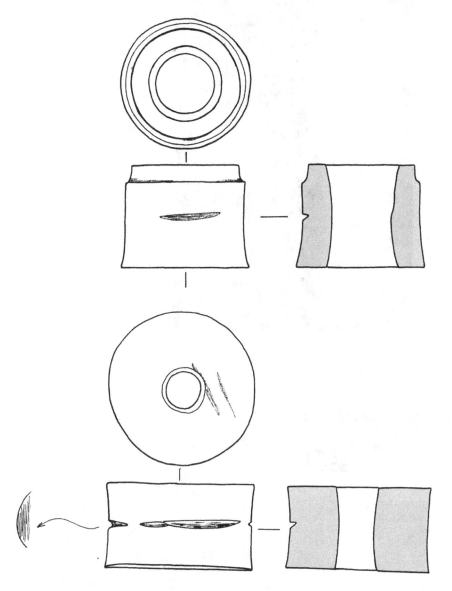

Figure 4.17 Blade-cut marks on the outer surface of the inward-curving *zhuxingqi* cylindrical object and *duanshi* end ornament from Yaoshan tomb no. M2

Figure 4.18 Blade-cutting technique used to create the string pattern on the dragon-head motif tube M2:7 from the Yaoshan site

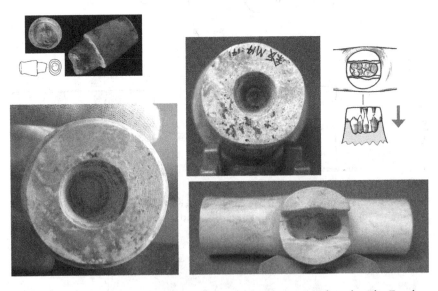

Figure 4.19 Technique used to shape the tenons and mortises found at the Fanshan and Yaoshan sites (lower right from Yaoshan M11:72); other three from Fanshan (M16:74-1, M14:141, and M12:101, top to bottom)

high-speed rotation technology used in fast-wheel throwing had become fully developed and easy to use. Unlike the potter's wheel that creates an upward force, the core drill applies a downward force.

Zhuxingqi cylindrical object M12:102 from the Fanshan site is 7.95 cm tall and slightly bulging in the middle, with a widest diameter of 4.2 cm. It was shaped with a tubular drill. Amongst the quartz *jue* slit rings from the jade workshop at Fangjiazhou site, some were shaped with abrasion and polishing, while others were made more directly by tubular drilling from one direction. The central hole of *cong* tube M12:98 from Fanshan, with an outer diameter of 5 cm and inner diameter of 3.8 cm, was also made by tubular drilling. It was very accurately drilled from both directions and the joint is barely mismatched.

The tenon of *duanshi* end ornament M16:74~1 from Fanshan was shaped by tubular drilling. *Duanshi* end ornament M12:101 from Fanshan and the mortise of *duanshi* end ornament M14:141 from Fanshan were both bored by tubular drilling (Figure 4.19).

The tubular drilling tool has a circular edge, whose rotation exerts a downward pressure and drives the abrasive sand. The largest tubular drill used for Liangzhu jade is the one used for *bi* disk M20:5 from Fanshan with a diameter of 10.5–10.6 cm. There are tubular-drill marks on its edge and densely packed concentric circular marks on the top surface (Figure 4.20). The smallest tubular-drill mark can be found on the suspension hole on the back of jade figurine 98M29:15 from the Lingjiatan site. A closer view shows that 'the left-side hole was made by tubular drilling three times and gradually moving towards the right. One of the cores from the drilling was left inside the hole and has a diameter of 0.15 mm' (Anhui 2006a, p. 248). It was previously thought that the holes of jade tubes were made with a solid drill. When I observed jade tube M18:3 from the Pishan site, I saw that there is a misaligned edge inside the hole, typical for tubular drilling (Figure 4.20; Zhangjiang 2006b, p. 42). During recent post-excavation work at the Xiaodouli site, we found that most tube holes were made by a tubular drill as well. While tubular drills with a long diameter were probably made with bamboo or similar materials, this type of mini tubular drills were most likely made with the limb bones of poultry and other types of birds. This adds important insights to our understanding of tools and mechanisms used for tubular drilling.

In addition to the size of the drills, another important aspect of tubular drilling is its depth. From what we know so far, the deepest tubular drilling was used to make *cong* tubes.

Another notable phenomenon is that concentric circular marks were left on the top and bottom surfaces of many cores. In addition to *bi* disk M20:5 from the Fanshan site mentioned previously, the jade core excavated from the Tangshan site also has small round indentations in the centre (Figure 4.21). Stone drill cores T0202②:4 and T0203⑥:13 (Zhejiang 2011, p. 197 and plate 73) from the Wenjiashan site in the Liangzhu Site Cluster and T0103④b:31, T0102④b:3, T0102④b:7, and T0103④b:18

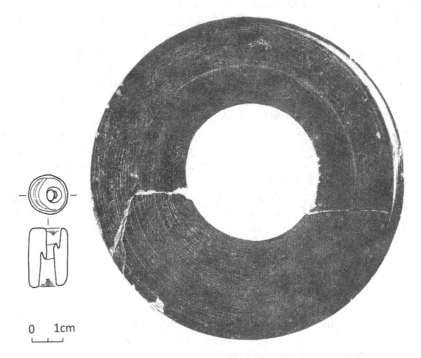

0   1cm

Figure 4.20 Jade tube M18:3 from the Pishan site (left) and jade *bi* M20:5 from the
          Fanshan site (right)

(Nanjing Museum 2001) from the Dingshadi site have similar marks.
Clearly, examples of this tubular-drilled mark are not rare. These concentric circular marks on tubular-drilled surfaces were caused by the friction
between the drilled object and the abrasive sand pressed down by the
drill. This shows that there was a *zhuxingqi* cylindrical object inside the
tubular drill exerting downward pressure, rotating with the tubular drill
and stabilising the drilled object, so that the tubular drill can sustain its
downward movement. A device like this is more complicated than the kind
of tubular drill simply made of a rotating tube that applies downward
pressure, as we might usually imagine.

A key question concerning tubular-drilling technology is whether the
drill or the drilled object rotated during the drilling process. In the latter
scenario, the drilled object should have been fixed on a fast wheel. In
the former scenario, it would require the use of a circular tubular drill
rotating at a high speed. The tubular-drilling technique used to make the
holes of jade tubes might be different from the technique used to make
yue axes, *bi* disks, and *cong* tubes. Perhaps both tubular-drilling methods
existed.

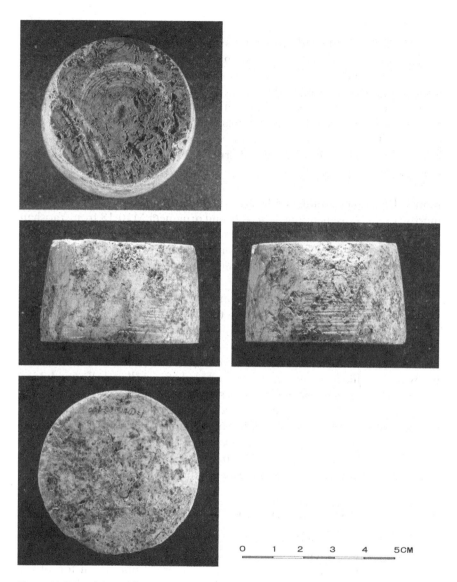

Figure 4.21 Jade item from the Tangshan location at the Liangzhu site cluster

It is noteworthy that concentric circular marks are also found on jade drill cores from the Lajia site in northwest China and on single-drill-direction granite cores from the Donglongshan site in Shangluo, dated to the Late Longshan period to Xia period (Gu 2005, p. 147; Shaanxi and Shangluo 2011, Figure 3.9). Clearly, the same tubular-drilling technique was used across a wide area over a long period after the Liangzhu Culture (Figure 4.22).

4.2.3 Boring

Tubular drilling and solid drilling were the two types of boring techniques used in jade production in the Liangzhu Culture. Boring refers to the making of mortises, which were the most numerous assembly parts with the most diverse shapes discovered in the Liangzhu site cluster.

As mentioned previously, tubular drilling was only used to make round mortises, whereas elongated, rectangular, and oval mortises required the use of borers. To make each drill hole with a consistent diameter when using a borer, it is necessary to use a conical drill bit and a drill rod in tandem. This is also seen as a primitive form of rotating wheels.[7]

There are at least seven solid drilling marks at the bottom of the oval mortise of the jade handle M11:72 from the Yaoshan site (Figure 4.19). The mortise of the aforementioned *duanshi* end ornament M10:18 from Yaoshan was made with twice of solid drillings. There are also numerous examples of mortises made using solid drilling only once. For example, mortises made this way are found on the top, bottom, and inner side of the jade handle M2:55 from Yaoshan.

The deepest result produced by Liangzhu solid drilling technique can be seen on the long jade tube M17:9 from the Fanshan site that was paired with the middle prong of a three-pronged object. The jade tube is 12.48 cm long and was solid drilled from two sides, one side reaching as deep as 11 cm. The solid driller used for this could not have been a handheld conical stone tool but was more likely a complex tool assembled on a drill rod (Figure 4.23).[8]

The holes in a type of long jade tubes from the Beinan site (1000–800 BC) in Taiwan (Zang 2005, p. 7) are also remarkably surprising. The jade tubes are more than 20 cm long, with either nearly square or round cross-sections less than 1 cm wide. They were very accurately drilled through from the two opposite ends (Ye 2005, pp. 27–29). To drill so deeply, this kind of solid drill should also be a combination of a drill bit and a rod.

4.3 Types, shapes, and carved motifs

Cong tubes, *bi* disks, and *yue* axes are representative jade objects from the Liangzhu Culture. In elite burials found within the Liangzhu site cluster and nearby sites, there are usually *zhuxingqi* cylindrical objects on top of the coffin and ornaments on both sides of the burials, but the majority of other jade objects are found near the head of the tomb's occupant, representing the identity, gender, and status of the person. Examples include *guanzhuangshi* cockscomb-shaped objects, sets of semi-circular ornaments, *sanchaxingqi* three-pronged objects that represent male prestige, and groups of *zhuixingqi* awl-shaped objects.

It is widely recognised that *cong* tubes, *bi* disks, and *yue* axes are the most important jade objects in the Taihu Lake region as well as important components of subsequent Chinese material culture. In fact, *cong* tubes and

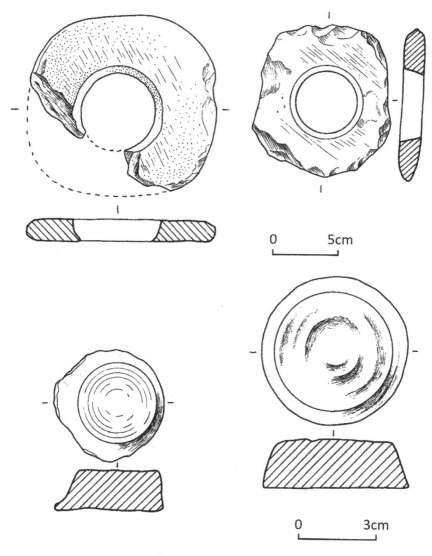

Figure 4.22 Marble *bi* and drill cores from the Longshan Culture site Donglongshan in Shangluo

bi disks were original creations of the Liangzhu Culture. Contemporary diffusion of *cong* tubes and *bi* disks reached the Huating site to the north and the Shixia site to the south, and spread along the Yangtze River to the areas along the Dabieshan Mountains. However, *cong* tubes and *bi* disks only reached the Middle Yangtze River and the Yellow River during the late Liangzhu period or the later Longshan period.

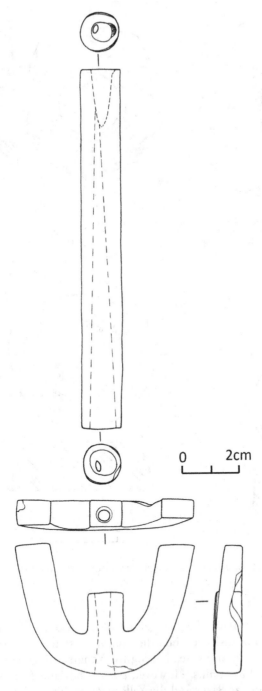

Figure 4.23 Long tube M17:9 that forms a set with *sanchaxingqi* three-pronged object, from the Fanshan site

The production of *cong* tubes was complicated. First, the raw jade must be quite sizable. An unfinished *cong* tube with guiding lines on both ends was found at the Wujiabu site in 1973. Its central hole had already been hollowed out (Figure 4.24) (Wang 2003, p. 88). Though there is a long-standing misunderstanding that the *cong* tube is round inside and square outside, in reality it is round inside but *not* square outside. Craftsmen intentionally gave a rounded shape to the outer surface of *cong* tubes probably to achieve special visual effects through its shape and decorative patterns (Fang 2013b). Regardless of height, all *cong* tubes are larger on top and smaller at the bottom to some extent. The central hole of all *cong* tubes were tubular drilled from two directions and then further adjusted and polished, except for tall *cong* tubes with multiple sections and small central holes. The protruding nose on the sacred human and animal-face motif on each section of the *cong* tube was made by reducing the background with repeated blade cuts. It would be even more complicated if large animal eyes needed to be carved and filled with intricate patterns. The division of sections on *cong* tubes also requires meticulously accurate calculations. The height of each section of the tall *cong* tube with multiple sections excavated from tomb no. M3 at the Sidun site have a margin of difference within 0.1 cm.

Due to their importance and uniqueness, *cong* tubes are seen as important symbols of settlement hierarchy and scale. Based on the distribution of *cong* tubes around the Taihu Lake region, Shin'ichi Nakamura divided the area into eight site clusters: Liangzhu site cluster, Tongxiang-Haining site cluster, Linping site cluster, Deqing site cluster, Haiyan-Pinghu site cluster, Wuxian-Pishan site cluster, Qingpu site cluster, and Changzhou site cluster (Figure 4.25) (Nakamura 2003). During the early and middle Liangzhu period, the vast majority of *cong* tubes were found within the Liangzhu site cluster. The Fanshan and Yaoshan sites, where a total of 40 *cong* tubes were discovered, far exceeded other sites from areas surrounding the Liangzhu site cluster and high-ranking settlements in southern Jiangsu Province, both in terms of quantity of *cong* tubes and quality of decorations. The similarities between *cong* tubes from the Gaochengdun site and from the Liangzhu site cluster show that there were close political ties alongside artistic interactions. During the late Liangzhu period (marked by the emergence of the bag-legged *gui* kettle), the number of tall *cong* tubes with multiple sections increased in southern Jiangsu and around Shanghai but decreased within the Liangzhu site cluster, showing that jade sources changed and technical preferences shifted during this period.

Yue axes, especially the complete ones with *mao* and *dui* end ornaments, also reflect settlement hierarchy. So far, *yue* axes with *mao* and *dui* end ornaments have only been found in a small number of high-ranking settlements such as the Liangzhu site cluster, the Fuquanshan site, the Caoxieshan site, and the Sidun site.

Sanchaxingqi three-pronged objects are mostly found in the Liangzhu site cluster and Linping site cluster, less so in the Tongxiang-Haining site

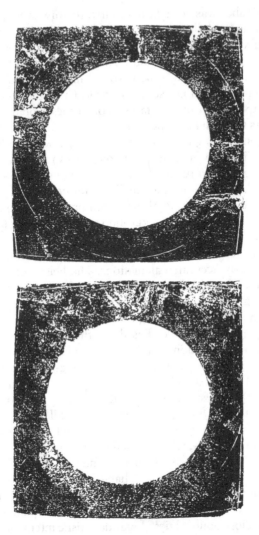

Figure 4.24 Rubbing showing the guiding lines on the top and bottom sides of the unfinished *cong* from the Wujiabu site

cluster, and none in southern Jiangsu and around Shanghai. *Sanchaxingqi* three-pronged objects are a very representative type of jade objects found in the Liangzhu site cluster and nearby sites. Those found in the Linping site cluster are of lower quality in comparison to those from the Liangzhu site cluster, while those from the Tongxiang-Haining site cluster are even more inferior. *Sanchaxingqi* three-pronged objects from the Tongxiang-Haining site cluster are rarely made with high-quality tremolite-nephrite but are usually made with pyrophyllite instead. Low-ranking settlements tend to have

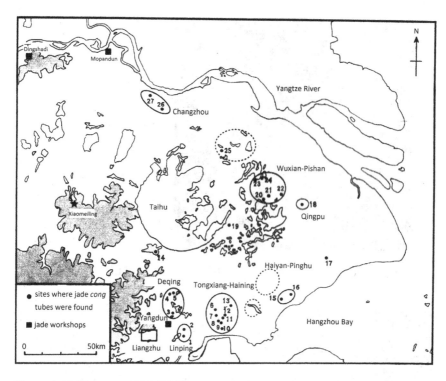

Figure 4.25 Nakamura Shin'ichi's grouping of Liangzhu Culture site clusters –
1. Hengshan; 2. Maoshan; 3. Xin'anqiao; 4. Huishan; 5. Zhuangqian; 6. Dianjietou;
7. Xindili; 8. Liyun; 9. Heyedi; 10. Shedunmiao; 11. Luowan; 12. Taozicun;
13. Pu'anqiao; 14. Yangjiabu; 15. Wangfen; 16. Daimudun; 17. Tinglin; 18. Fuquanshan;
19. Wangyancun; 20. Zhanglingshan; 21. Zhaolingshan; 22. Shaoqingshan; 23.
Caoxieshan; 24. Chuodun; 25. Jialingdang; 26. Sidun; 27. Gaochengdun

none or only a single three-pronged object at each site. In any case, very
few three-pronged objects have been found in this type of settlements. For
example, only three such objects have been found among seven burials in
the late Liangzhu period cemetery at the Yaojiashan site (Wang et al. 2009,
pp. 88–90). Though *cong* tubes, *bi* disks, long *zuixingqi* awl-shaped objects,
and *guanzhuangshi* cockscomb-shaped objects were found in 140 Liangzhu
period burials at the Xindili site, only one three-pronged object (M109:5,
made with pyrophyllite) was present (Figure 4.26).

The relationship between the Liangzhu site cluster and surrounding sites
was fixed since the early Liangzhu period. The ceramic filter, which is an
object with assembled parts and is associated with the female gender, is a
typical example. Mainly distributed within the Liangzhu site cluster during
the early Liangzhu period, it was found in ordinary burials at the Miaoqian
site as well as in tomb no. M23 at the Fanshan site. Recent archaeological

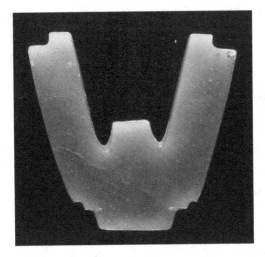

Figure 4.26 Sanchaxingqi three-pronged object M108:5 from the Xindili site

Figure 4.27 Ceramic filter from burial (left) and pit fill (right) at the Wujiabu site

discoveries show that this type of ceramic vessel has also been found at the Guanjingtou site in the southern part of the Liangzhu site cluster, and at the Yujiashan site in the Linping site cluster, but is absent from the Jiaxing area, in southern Jiangsu and around Shanghai (Figure 4.27) (National Museum of China and Zhejiang 2005, pp. 70–71).

Few *guanzhuangshi* cockscomb-shaped objects have been found in southern Jiangsu. One *guanzhuangshi* cockscomb-shaped object with a special shape has been excavated from tomb no. M199 in Caoxieshan site. It has a protruding part at the back and a hole perforating the protruding part. A *guanzhuangshi* cockscomb-shaped object from the Zanmiao site, the westernmost site where *guanzhuangshi* cockscomb-shaped objects have been excavated, has animal-face motifs carved on the front and back. There is also a perforated protruding part above each large animal eye on the back. This type of *guanzhuangshi* cockscomb-shaped object with a protrusion on the backside seems to be a hybrid between *guanzhuangshi* cockscomb-shaped objects and *sanchaxingqi* three-pronged objects (Zhejiang et al. 1990, Figures 113, 114, 116, and 117).

The shape and decoration of *cong* tubes from the previously mentioned Gaochengdun site reveals close connections to those from the Liangzhu site cluster, but no *guanzhuangshi* cockscomb-shaped object has yet been discovered in the 13 elite burials at Gaochengdun, though they are common in the Liangzhu site cluster.

If one compares jade types from the Liangzhu site cluster with those found at surrounding sites, it is obvious that whatever is found within the Liangzhu site cluster tend to be absent at the surrounding sites. When it is present non-tremolite-nephrite stones are used instead. Some small jade objects not seen within the Liangzhu site cluster were excavated at the Fuquanshan and Qiuchengdun sites, but their occurrence is limited to individual tombs and mainly date to the late Liangzhu period. Examples include jade handle M144:25 from the Fuquanshan site (Shanghai 2000, p. 84) and jade tube pendant CJ:5 with human-face motif from the Warring States period (476–221 BC) earthen mound at Qiuchengdun site (it is probably an unusual *zhuixingqi* awl-shaped object that functioned as an insert part) (Jiangsu 2010, p. 220) (Figure 4.28).

As for the sophisticated sets of *duanshi* end ornaments, strands of jade tubes and semi-circular tubes, and crescent-shaped ornaments found in the Liangzhu site cluster, surrounding sites have nothing comparable (Figure 4.29).

Just like how the type and shape of jade are representative of differences in rank, resource, and craftsmanship between settlements, decoration (especially those associated with certain object shapes) is also representative of these differences. Dragon-head motif and sacred human and animal-face motif are the two major types of ornamentation on Liangzhu jade objects. *Cong* tubes are always decorated with the mask motif, hence the idea that the mask motif can be considered the soul of *cong* tubes.

The dragon-head motif developed from single jade dragons in the late Songze period. It took the figure of the dragon's head and evolved it into a continuous dragon-head pattern. There is a close connection between the dragon-head motif and the animal-mask motif. The dragon-head motif disappeared from jade objects after the early Liangzhu period. Based on jade objects with dragon-head motif found to date, they mostly appeared in the Liangzhu site cluster and at surrounding sites, though stand-alone

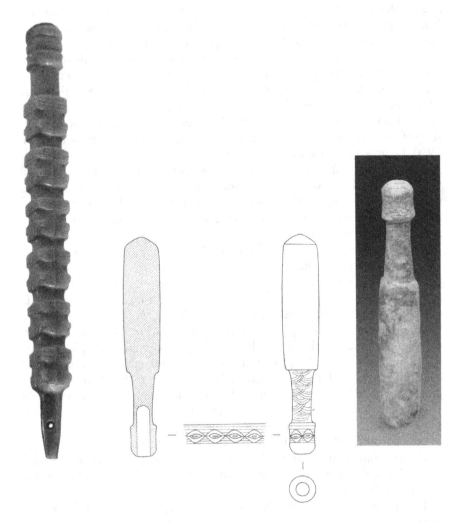

Figure 4.28 Relevant jade objects from the Fuquanshan (left) and Qiuchengdun (right) sites

Liangzhu jade dragon figurines were also discovered in Southern Jiangsu and the Hangzhou-Jiaxing-Huzhou Plains (Figure 4.30). Dragon figurines and mini circular dragons with a gap have been excavated from the Puanqiao site. The mini dragons are only 0.7 ×1 cm big (Zhejiang and Liangzhu Museum 2014).

Most jade objects with the dragon-head motif have been found in the Liangzhu site cluster, which shows that this region already led opinions on the conceptual and visual aspects of jade objects since the early Liangzhu period (Figure 4.31).

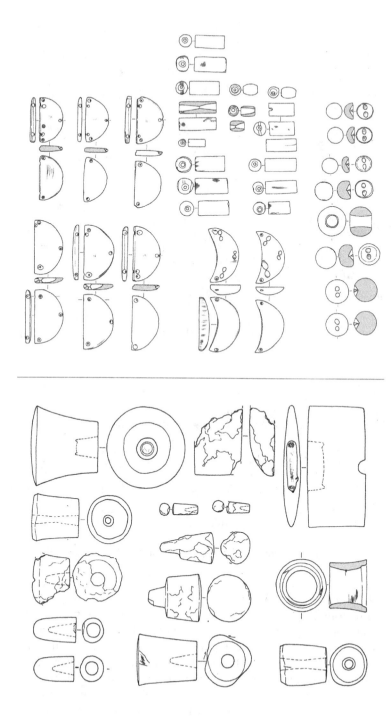

Figure 4.29 Assorted finial ornaments from Fanshan tomb no. M14 (left) and assorted tube pendants and other pendants from Yaoshan tomb no. M10 (right)

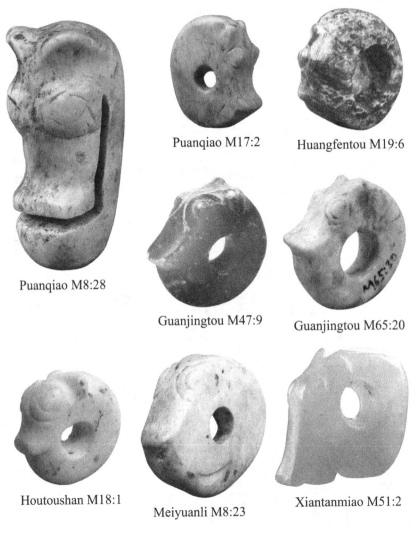

Puanqiao M17:2　　　Huangfentou M19:6

Puanqiao M8:28

Guanjingtou M47:9　　　Guanjingtou M65:20

Houtoushan M18:1

Meiyuanli M8:23

Xiantanmiao M51:2

Figure 4.30 Jade dragons from the Liangzhu site cluster and the surrounding area

Sacred human and animal-face motif (including the simpler 'animal face') can be found in the form of a simple two-dimensional motif, a three-dimensional motif on the corner of *cong* jade tubes, and occasionally, a diagonal three-dimensional motif. Complete sacred human and animal-face motifs had already been found before the excavation of the Fanshan site, but people did not pay it much attention or link it to the decoration on *cong* jade tubes. In the preliminary report of tomb no. M3 at the Sidun site in the second issue of *Kaogu* in 1984, the decorations on *cong* tube M3:4 were described as 'symbolic animal-face motif' and 'realistic animal-face

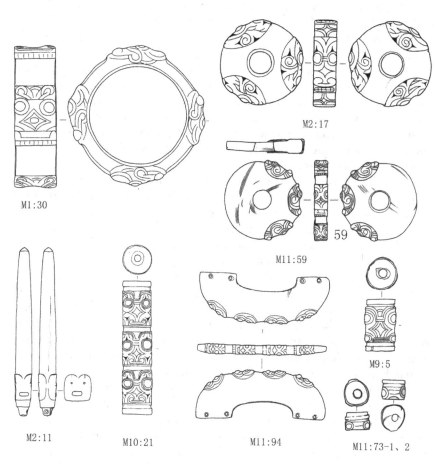

M2:17

M1:30

M11:59

59

M2:11

M10:21

M11:94

M9:5

M11:73-1、2

Figure 4.31 Jade objects with the dragon-head motif from the Yaoshan site

motif'. Though the raised horizontal parts are noses, they were interpreted as 'mouths'. In 1986, Deng Shuping referred to the decorations on *cong* tubes as 'mask motif with small eyes' and 'mask motif with large eyes' (Deng 1986). The excavation of the Fanshan and Yaoshan sites finally clarified the relationship between the mask motif and the decorations on *cong* tubes, as well as deepening our understanding of the components of the mask motif. The preliminary report of the Fanshan site referred to the decorations on the vertical grooves and corners of *cong* tube M12:98 as the '*shenhui* insignia' and 'simplified *shenhui* insignia'.

Forty *cong* tubes have been excavated from the cemeteries at Fanshan and Yaoshan, among which there are 23 occurrences of two-dimensional sacred human and animal-face motif and 24 'animal-face' motifs.

Apart from objects from the Fanshan and Yaoshan sites, the mask motif has only been seen on ivory sceptres discovered at the Fuquanshan and Wujiachang sites. The largest number of objects with sacred human and animal-face motifs was excavated from tomb no. M12 at the Fanshan site. The eight complete mask motifs on the vertical grooves of *cong* tube M12:98 from Fanshan can be considered archetypal examples of this motif. The criteria for an 'archetypal' example include the completeness of major components of the motif and highly symmetrical and repetitive use of the motif on the same object. The sacred human and animal-face motif is composed of a crouching divine animal and the upper body of the sacred man. The face and headgear of the sacred man, and the eyes, nose, and mouth of the sacred monster are blade-cut bas-relief, and the rest are expressed by incised lines (Figure 4.32).

The animal-face motifs on objects M9:1, M9:2, *zhuxingqi* cylindrical object M11:64, and *zhuoshicong* bracelet-style *cong* tube M9:4 from the Yaoshan site are carved inside rectangular friezes that fit the shape of the objects. The sacred human and animal-face motif and animal-face motif on *guanzhuangshi* cockscomb-shaped object M11:86 from Yaoshan, M15:7 from Fanshan, and M16:4 from Fanshan adapt the motif to the ridged top of the object. Similarly, the animal-face motifs on three-pronged object M9:2 and M3:3 from Yaoshan are also fitted to the shape of the object (Figure 4.33).

Ornamentation on *cong* tubes are the best examples of motifs adapted to the shape of the object. By distinguishing between the sacred human and animal based on the differences between their eyes (size and presence of the crescent-shaped ear), we can divide the motif into four types: single sacred human, repeated sacred human, sacred human and animal face, and repeated sacred human and animal face (Figure 4.34). Mou Yongkang argued that roles of 'man' and 'animal' change over time; initially there was a close connection between the man and the animal, then a certain distance emerged between the two, then finally the motif was simplified by omitting any distinction between man and animal. Such changes 'demonstrate how the worship of anthropomorphic god was the driving force behind this change' (Mou 1989b, p. 190). Though a simplified style had existed since the Early Liangzhu period, and the animal-face motif was still in use during the late Liangzhu period as seen on *zhuoshicong* bracelet-style *cong* tube M3:43 from the Sidun site, it is also true that the animal-face motif is not seen on tall *cong* with multiple sections dated to the late Liangzhu period.

Burial no. M200 at the Yujiashan site is an elite woman's burial with abundant jade finds, including openwork *guanzhuangshi* cockscomb-shaped objects, *huang* semi-circular ornaments, *huan* ring bracelets, *zhuoshicong* bracelet-style *cong* tubes, dagger-like objects, and chopstick-shaped objects. The four animal-face motifs on *zhuoshicong* bracelet-style *cong* tube M200:58 seem to have the same components and composition as examples from the Yaoshan site, but the details are quite different when examined closely. For example, the lines are stiff. The upward-slanting, pointy beak-like decoration (*jianhui*) of the large animal eye is even carved outside of the raised eye contour. There is not a single spiral among the lines that fill

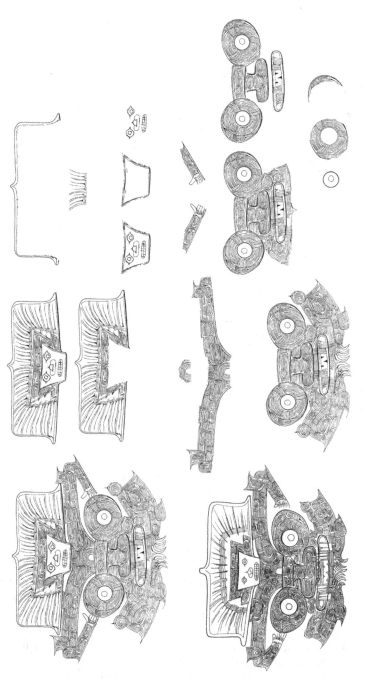

Figure 4.32 Components of the complete sacred human and animal-face motif on the *cong* tube (top left) and *yue* axe (bottom left) from Fanshan tomb no. M12

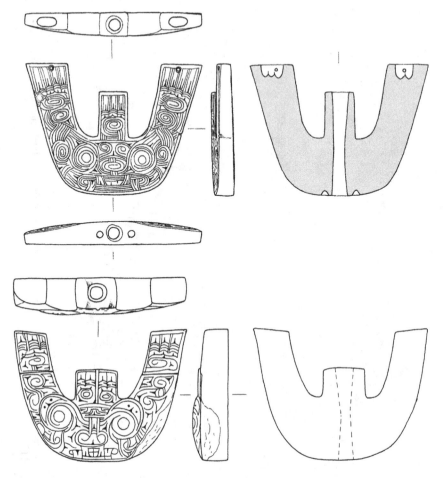

Figure 4.33 Sanchaxingqi three-pronged object from the Yaoshan site (top, tomb no. M9; bottom, tomb no. M3)

the outer circle of the large animal eyes. The radial feather-like lines on the raised part between the eyes are satisfactory on one of the four motifs, but the other three completely miss the point. Just like the spirals, the supposedly beak-like pointy decorations of the eye do not have a beak-like shape at all. Although trace element analysis of these jade objects shows that they were made with similar raw material as jade objects from the Fanshan and Yaoshan sites, they were clearly made by different craftsmen. Thus, these are representative examples of shoddy copies of Liangzhu imagery (Figure 4.35) (Zhejiang Cultural Heritage Bureau 2001, p. 170).[9]

There is an ivory sceptre about 97 cm long from tomb no. M207 at the Wujiachang cemetery (Shanghai Museum 2010), decorated with sacred

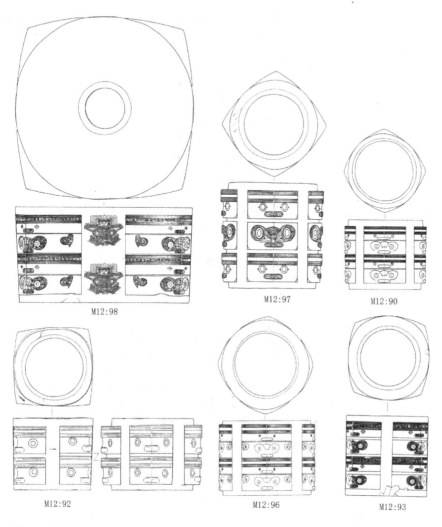

Figure 4.34 Decoration on the sections of *cong* tubes from Fanshan tomb no. M12

human and animal-face motif and complete with a mounted ivory *dun* ornament. The sceptre was made with thinly cut sheets of ivory. One side has
a total of ten bas-relief sacred human and animal-face motifs, made with
the 'reducing background' (*jiandi*) blade-cutting technique. Two bird motifs
and two animal-face motifs were carved on the *dun* ornament of this staff
(Figure 4.36) (Shanghai Museum 2014, pp. 132–134).

The discovery of this elaborate ivory sceptre at the Wujiachang site
solved questions concerning the shape and function of the ivory object
from Fanshan tomb no. M20 and lent insights to the reconstruction of

Figure 4.35 Shoddy mistakes of the animal-face motif on *zhuoshicong* bracelet-style *cong* tube M200:58 from the Yujiashan site

the elaborate sceptre from Fanshan tomb no. M12. The *mao* and *dui* end ornaments of the elaborate sceptre from Fanshan tomb no. M12 are closely connected to the ivory sceptre from Wujiachang tomb no. M207 in terms of both shape and structure. Also, *cong* tube M12:90 was probably fitted onto the ivory sceptre from Fanshan tomb no. M12 (Figure 4.37).

There were nine ivory objects on top of the coffin in tomb no. M20 at Fanshan. Due to the bad state of preservation, the excavation report once interpreted that 'one end was cut to be worked into a tip' (Zhejiang 2005a, p. 221). Now it seems that the parts can be separated, one of them being a *dun* ornament that can be mounted. However, the ivory objects were not cut into thin flat shapes. The end of ivory sceptre M20:9 has a diameter of 8 cm, but the body of the sceptre might be even narrower. One ivory sceptre was even fitted inside the central hole of *cong* M20:124, which has an inner diameter of 6.5 cm (Figure 4.38).

Tomb no. M12 at the Fanshan site and tomb no. M207 at the Wujiachang site, respectively, representing the beginning and the end of Liangzhu Culture, show that the sacred human and animal-face motif maintained the same overall appearance and elaborate sceptres kept the same shape. However, there are significant changes to the details of the decorative pattern. Lines became sparser and stiffer, while the curvature of spirals also became more awkward. The crescent-shaped ears beside the large animal eyes became more pointy. Just like the 'peach-shaped eyes' (a peach-like overall shape of the animal eye, with the round eye and the above-mentioned pointy 'ears' put together), there were new additions to the lines that filled the headgear in the mask motif. Bird motifs looked identical to the ones found on contemporary ceramic vessels and were very different from the imagery of a bird's body with animal face and large eyes as seen on objects from the Fanshan and Yaoshan sites during the early Liangzhu period. A cocked 'tail' was also added to the bird motif, just like the 'sideview sacred insignia' and 'sacred insignia' motifs

Figure 4.36 Ivory sceptre M207:61 from the Wujiachang site

on the *dou* stemmed dish and *hu* pitcher found at the Putaofan site in the Liangzhu centre, and the patterns decorating three-pronged object M28:4 from the Yujiashan site and the *shuangerhu* two-eared jar M198:7 from the Caoxieshan site. All these show that the overall appearance of the sacred human and animal-face motif did not change in the late Liangzhu period, but the details of the pattern had developed and changed.

The sacred human and animal-face motif seen on objects from early Liangzhu sites Fanshan and Yaoshan not only constituted the basic element of decorative patterns but also demonstrated the use of the most elaborate techniques, including micro-carving and blade cutting to shape bas-relief (*fudiao*) and to create grooves (*dawa*). As shown by measurements of decorations on the *mao* ornament on the elaborate sceptre M12:103 from

Figure 4.37 The elaborate sceptre from Fanshan tomb no. M12, possibly once fitted
  through a *cong*

Fanshan, at least three lines can be carved within 1 mm without overlapping
(Figure 4.39). This truly deserves the name of miniature carving.

There are six groups of spiralling, interwoven, and alternating sacred
human and animal-face motifs and animal-face motifs each on the *zhuxingqi*
cylindrical object M12:87 from Fanshan. The sacred human and animal-
face motifs are 2.7 cm by 3.5 cm each if spread out flat, with animal eyes
measuring only 0.6 × 0.7 cm. Within such a small space, strands of non-
overlapping fine lines fill the circles of the eyes, tiny beak-shaped patterns
and spirals fill the crescent-shape ears, and grooves were not omitted either
(Figure 4.40). These intricate patterns not only deliberately showcased the
fine craftsmanship but also created imagery far more vivid than people's
normal visual perception.

Figure 4.38 Ivory objects excavated from Fanshan tomb no. M20

Figure 4.39 Intricate patterns on the *mao* end ornament of the elaborate sceptre
M12:103 from the Fanshan site

Figure 4.40 Sunken grooves and decorations on the large eye of the sacred man and animal-face motif of *zhuxingqi* cylindrical object M12:87 from Fanshan

From the middle Liangzhu period onwards, this kind of minutely carved jade objects gradually disappeared and lines became awkward and stiff, showing that there was a change in the control of the spiritual realm.

4.4 Discussion and summary

Ivory, lacquerware, silk, and especially jade represented cultural conceptions and beliefs of the Liangzhu Culture and can reflect the rank and scale of settlements and identity and status of tomb occupants. With the establishment of male dominance during the late Songze period, these objects also became symbols of male power. The quality and quantity of jade as well as their type, grouping, shape, and pattern became the sole indicator of differences in rank and scale of each site cluster. The Liangzhu site cluster had a head start from the beginning and clearly established a leading position in the surrounding area since the late Songze to early Liangzhu period. The hydraulic system (see Chapter 2), used for flood prevention and transport, in the hills north-west of the Liangzhu site cluster has been 14C dated to 3100–2900 BC, showing that elaborate long-term planning for this site had been conceptualised and carried out very early on.

From the 12 sites excavated by Shi Xingeng in 1936 including Liangzhu, Changming, and Anxi, the notion of a Liangzhu site cluster encompassing nearly 4,000 ha and 45 sites proposed by Wang Mingda in 1986, to the 135 sites spanning more than 4,200 ha detailed in the archaeological report *Liangzhu site cluster*, our understanding of the complex Liangzhu site cluster has changed significantly throughout the years. With further archaeological investigations of the urban centre of the Liangzhu site cluster, the discovery of the hydraulic system, and the establishment of the notion of a Daxiongshan site cluster, south of the Liangzhu centre, it is not difficult to imagine the Liangzhu site cluster occupying an even more extensive area directly controlled by the walled city in the Liangzhu site cluster. Based on the distribution of ceramic sieves and jade objects excavated from elite tombs at the Hengshan and Yujiashan sites, the Linping site cluster appears to have the closest connection with the Liangzhu site cluster.

The Tongxiang-Haining site cluster occupies the area between Tongxiang, Haining, and Haiyan. It is at some distance from the Linping site cluster to the west and borders on the Qiantang River to the south. The land to the north is low-lying where the sites are sparsely scattered, especially further north towards districts like Jiashan. Though there are important sites such as Zhuangqiaofen and Tinglin to the north-east, they seem to transition to the high-ranking Fuquanshan site. The Tongxiang-Haining site cluster is a rather independent group between the late Songze period and Liangzhu period. Main sites that have been excavated include Dazemiao, Heyedi, Shedunmiao, Xiaodouli, and Huangfentou in Haining City; Puanqiao, Xindili, and Yaojiashan in Tongxiang City; and Longtangang and Xiantanmiao in Haiyan City. *Cong* tubes are found at many sites, but carved jade objects are rare. The jade also seems to have come from diverse sources. The jade assemblage found in the early to middle Liangzhu period tomb no. M6 at the Xiaodouli site is a representative example of jade used in the Tongxiang-Haining site cluster during this period. There is tremolite-nephrite with a 'chicken bone white' colour and some with an elegant green colour as well as an abundance of phyrophillite objects (Figure 4.41).

As for the Songze and Fuquanshan sites in the Qingpu site cluster, the former was continuously occupied since the Songze period, while the latter was founded in the late Songze period and maintained a continuous sequence throughout the Liangzhu period. During the Songze period, the settlement rank of Songze was higher than Fuquanshan, but the latter developed

Figure 4.41 Jade objects from Xiaodouli tomb no. M6

rapidly from the onset of the Liangzhup. During the late Liangzhu period, Fuquanshan and Wujiachang sites became comparable and complimentary high-ranking settlements. One of the late Liangzhu elite tombs at Fuquanshan had an abundance of large jade axes, some exceeding 0.3 m in length. It also had quite a high number of *bi* disks compared to the Liangzhu site cluster and surrounding areas. Seven vertically arranged *bi* disks were found inside the coffin and one at each end of the coffin in tomb no. M204 at Wujiachang (Zhou 2010, p. 4). The practice of an orderly vertical arrangement of jade *bi* disks beside the tomb's occupant was also seen in tomb no. M5 at the Qiuchengdun site but was not seen in the Liangzhu, Linping, and Tongxiang-Haining site clusters. Though many *bi* disks have been found at tomb no.

M4 at the Houyangcun site in the Liangzhu site cluster, they were randomly placed near the feet of the deceased, just like at Fanshan. Joint burials were found at Fuquanshan and Wujiachang. Tomb no. M139 at Fuquanshan even showed traces of possible sacrificial burial (Shanghai 2000). Jade quality and colour at Fuquanshan and Wujiachang, especially during the late Liangzhu period, were clearly different from the Liangzhu site cluster. Their burial practice was also very distinctive, especially burials at Fuquanshan. Site construction and occupation seem to be built mainly vertically, layer upon layer, at these two sites. All these facts show that the Qingpu site cluster had an equivocal relationship to the Liangzhu site cluster.

The Qingpu site cluster and its surrounding areas had no jade sources at all. Qin Ling had noticed some important facts quite early on. During the later section of layer two at Fuquanshan site (equivalent to the late Liangzhu period), the ability to amass socio-economic resources at Fuquanshan increased rapidly. All types of luxury items obtainable outside of the Liangzhu site cluster appeared here. There is clear evidence of long-distance interaction with the Shandong area. These phenomena show that Fuquanshan consolidated its position as the centre of this area and expanded its sphere of influence, vastly increasing its influence over the entire Taihu Lake region. Also, '[c]entrally located in the Taihu Lake region without any jade source nearby or a local tradition in jade production, Fuquanshan can only obtain luxury objects symbolising prestige through various different channels to display its social power' (Qin 2005, pp. 33, 35).

The Wuxian-Kunshan site cluster along the banks of Wusong River, including the Caoxieshan, Zhanglingshan, Shaoqingshan, and Zhaolingshan sites, flourished mostly during the early to middle Liangzhu period.

Located between the Taihu Lake region and the Ningzhen Plains and close to the Yangtze River, the Changzhou site cluster that encompasses sites, such as Gaochengdun and Sidun, had a distinctive culture since the Songze period. The Xingang site reflects a mix of characteristics from the Taihu Lake region, the Ningzhen region, and other areas along the Yangtze River (Changzhou Museum 2012). The Gaochengdun site did not maintain its close connection to the Liangzhu site cluster for long during the early to middle Liangzhu period before the period represented by the tomb no. M3 at Sidun, representative of elite culture in the Changzhou site cluster, began developing its own distinctive use of jade objects. On the one hand, the basic combination of *cong* tubes, *bi* disks, and *yue axes* is seen here; the shape and placement of large *cong* tubes are the same as tomb no. M12 at Fanshan; and *zhuoshicong* bracelet-style *cong* tubes were also worn here. On the other hand, there is an extremely high number of tall *cong* with multiple odd-number sections arranged in an orderly manner. Diverse jade colours also point to new jade sources.

Overall, the fine craft of jade production and the systematic use of jade remained similar throughout the Taihu Lake region during the Liangzhu period. There were no major changes to the sacred human and animal-face motif and the shape of the *cong* tube. Jade *yue* axes continued to be the symbol of male prestige. Quality tremolite-nephrite was the primary material for assemblages. The Linping site cluster showed the closest connection to the Liangzhu site cluster, followed by the Tongxiang-Haining site cluster, but the Qingpu site cluster and Changzhou site cluster both pursued their own development paths. Perhaps the differences in resources, location, and cultural continuity were precisely the causes for crises lurking behind the flourishing of this controlled fine craft.

Notes

1 According to Wen's study, the decomposition of jade undergoes six stages of changes on its transparency, color, and surface preservation. '*Lie qin*' represents the most severe stage of jade decomposition (Wen 1994).
2 'Mountain stream' gravel jade refers to weathered stones transported through the river water to the upper stream of rivers, normally with sub-angular shape and shiny surface. 'River-polished' jade refers to well-rounded stones after long-distance transportation in the river, normally of variable sizes and having a reddish-brownish surface (Gu and Li 2009, pp. 77 and 95).
3 '*Xiansuo*' refers to using flexible strings to cut jade with sand. This way of cutting could also help shape the crude forms of jade before further finer crafting works.
4 Incomplete large jade *zhang* tablets K1:81 and K1:97 from the sacrificial pit no. 1 at the Sanxingdui site reach 162 cm in length and 22–22.5 cm in width, both only 1.8 cm thick (Sichuan 1999, p. 63).
5 *Dawa* refers to a procedure using short-blade tools to chisel and polish the surface back and forth to create the relief.
6 Deng Cong believes this is secondary processing using the string-cutting technique, which is also a sound argument (Deng 2005, p. 77).
7 Such blades are thick in the middle and thin on the edge. They are fixed on a wheel and used to polish jade through rotation.
8 Assembled drill sets had been found in the Hemudu Culture at the Tianluoshan site in Yuyao and at Cihu site in Ningbo (e.g., Zhejiang and Ningbo 1993, pp. 108–109 and Figure 9.1).
9 The author would like to thank Mr. Lou Hang, director of the excavation of Yujiashan, for providing the photographs.

5 From the 'Songze Style' to the 'Liangzhu Mode'

Zhao Hui, translation by Wang Shaohan

5.1 Problems and methods

Since the 1980s, important, successive archaeological discoveries have challenged scholars' attitudes towards the Liangzhu Culture and the associated period. Ever since, research on the Liangzhu Culture has had two major focuses: Some scholars considering the Liangzhu Culture as representing the peak of socio-economic development of prehistorical societies across China during this period (what I call an 'evolutionary approach'), whilst other researchers are attracted by the specific characteristics of the rich material cultural relics from Liangzhu. These scholars conjecture that the social operation behind these unique material expressions may differ from those of other societies in that era (what I would term a 'case study' or 'historicism approach') (Zhao 2000, 2003). The author is among the minority who focuses on the latter approach. I have simply summarised the specificities of the Liangzhu Culture. The most remarkable aspect is the strong religious atmosphere presented by the carved motifs on the jade objects and pottery (Zhao 1999). The original plan was for the author to undertake more detailed investigation on these objects; unfortunately, that plan had been delayed until now. It has to be acknowledged that religions, ideologies, and related questions in ancient societies are fascinating topics of study. However, as pointed out by various textbooks, the issue of only using material evidence to study these questions, especially for prehistoric periods which do not have any textual records, is much more difficult, particularly when compared with the tasks of investigating ancient economic and technologies or reconstructing social organisations.

Fortunately, things have not gone to this extreme. Humans, who have the same social life, share the same thoughts, even the same way of thinking. As Renfrew stated, they share a mutual cognitive map (Renfrew and Bahn 2004). Based on this cognitive map, they create numerous things which are the same or similar. In other words, the similarities of those cultural relics could be regarded as an expression of people's mindset. They physically represent people's collective consciousness by means of material resources. In this sense, the Liangzhu material cultural traits could be used as a guide to explore their collective consciousness.

Nevertheless, there should be limitations to the extent for such explorations. Regardless if it is the collection of similarities in material resources or a mindset, in fact, it includes people's cognition on all aspects of societal and environmental issues arising from their time including astronomy, geography, and cosmology as well as craft production and stylised ideas inscribed on products. All aspects of these patterns of thinking are contained and/or inscribed onto material remains to various degrees. Discussing these issues in detail from all the aspects listed previously is impractical. Thus, this chapter will focus on and explore the social beliefs of the Liangzhu people through an examination of the material evidence of the carved motifs and morphological characteristics of selected objects.

In the archaeological investigations of prehistoric ideologies, we need to define some of the archaeological phenomena such as symbols. However, this definition is based upon our present knowledge and understanding, which means that its veracity cannot be proved. To avoid misinterpretation, misrepresentation, and oversimplification, I will do the following in this chapter:

1   While judging the nature and properties of certain archaeological aspects, I will start from the everyday life of the Liangzhu people. If no answer can be reached from such perspectives as utilitarian function and technologies, then issues about beliefs, religions, and sacrifices will be considered. For example, pits found at the Longshan Culture sites in Shandong Province are called 'sacrificial pits' by some scholars as a large amount of pottery sherds are found in these pits. However, we need to understand that the ceramic industry was highly developed in the Shandong Longshan Culture. It is a common phenomenon at these sites that ceramic products with slight defects were discarded at this time. Thus, to estimate the nature of these remains, the taphonomical issues and the relationship between the pit and its surroundings need to be considered. Only when all possible explanations arising from activities of daily life are excluded will it be possibly to say that these types of archaeological remains are related to sacrificial offerings.

2   Avoiding interpreting the meanings of archaeological objects based solely on historical records. Although the *cong* tubes and *bi* disks from the Liangzhu Culture are not for daily use, the question remains whether they were really used as the religious instruments offering sacrifices to the Heaven and Earth, as *Rites of Zhou* recorded (Zheng 2010)? As *Rites of Zhou* was written 2,000 years later than the Liangzhu Culture, this text can only be used as a reference to one of the many possibilities of the symbolic and social meanings of the Liangzhu jade. Some researchers interpret the sacred insignia with the human face and animal body, a popular decorated jade item during the Liangzhu period (see Chapter 4), as 'Three Qiao' (*San qiao*, the general term for the three achievement methods in Taoism). They then endowed the religious meaning of 'Three

Qiao' to the ideology of the Liangzhu people (Zhang 1986, 1988). In this author's opinion, that link is farfetched.

3   This chapter seeks explanations from common and recurrent motifs and patterns that can be seen on many different types of objects. This kind of explanation has comparative social significance, which is what this chapter attempts to reconstruct.

I have been thinking about this topic for a long time but have been too busy to carry out any research. In 2009, Sheng Qixin from Taiwan came to Peking University and became a PhD student of mine. Following my suggestion, she chose the decorative motif system of the Songze Culture and the Liangzhu Culture as her research topic. In her doctoral dissertation, she did an excellent detailed analysis of numerous decorative motifs from the Songze and the Liangzhu Culture and accurately summarised their characteristics (Sheng 2014). Thanks to her dissertation research, which provided a good starting point for my own analysis. Hence, this chapter does not need to start from the beginning to analyse the motifs on a large number of objects; rather, my research has directly benefitted from her analysis of numerous Song and Liangzhu decorative motifs

5.2 The Songze style

It is often best to situate objects or things in their boarder, historical contexts. In this sense, we need to look at the Songze Culture, the predecessor of the Liangzhu, to better understand the Liangzhu Culture. Unfortunately, few material remains have been recovered from the post-Liangzhu period. Therefore, this discussion will only focus on the origins of the Liangzhu Culture and not discuss its later developments.

The Songze Culture here refers to the Greater Songze region, which covers the area across the western part of the Yangzi River in Anhui Province to the East China Sea coast and from the central Jiangsu Province to the Qiantang River. It includes a group of archaeological cultures including Songze, Longqiuzhuang, Beiyinyangying, Lingjiatan, and Xuejiagang (Figure 1.2). By extending the geographical scope, it should be easier to find that the differences among these archaeological cultures are less than the differences between these cultures and their neighbours. Thus, we can regard these archaeological cultures as a Songze Culture Group.[1]

The largest number of archaeological remains discovered from the Songze Culture are ceramics. Because potters in that time usually put more creativity in the ceramic manufacturing processes, we do find objects in human or animal form were produced (Figure 5.1, nos. 5 to 10 and 14). Some of those objects are representational, some are exaggerations, and some are even humorous with cartoon-like characteristics. However, such ceramics with artistic patterns and designs are limited in numbers. Besides the *zhuxinghe* pig-shaped vessels (varying greatly in size) found at Longqiuzhuang, no

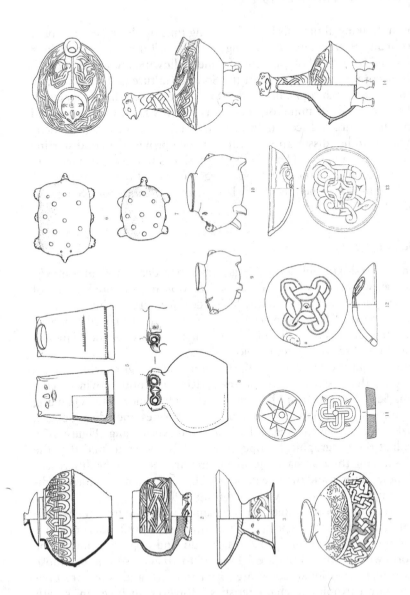

Figure 5.1 The pottery ornamentations and pictographic potteries in the Songze Culture – 1. Songze M59:2; 2. Nanhebang M59:1; 3. Zhili M9:1; 4. Shimadou M46:2; 5. Nanhebang M59:22; 6. and 7. Nanhebang M27:14 and 15; 8. Nanhebang M15:2; 9. Longqiuzhuang M383:4; 10. Longqiuzhuang M157:1; 11. Lingjiatan 98M19-16; 12. Xiaodouli M46:2; 13. Zhaolingshan M51-1:9; 14. Collection of the Jiaxing Museum

same-style products have been found. It would appear that the potters in a free-style manner created these objects.

The phenomenon of improvisation is rarely found in the shapes and decorations of daily-use ceramic objects. Generally, the majority of Songze ceramics are with no or little decorations. No graphic patterns can be recognised among the commonly seen decorations, including cord, addition, string, and press patterns. Only from the compositions of carved patterns, and engraved or incised patterns without hollow carving, can we attempt to glimpse the ideas of the prehistoric potters and/or decorators. Engraved or incised patterns without hollows were usually decorated on the handle of *dou*-stemmed dishes during the early Songzhe period. The majority of these patterns are of rectangular and triangular shapes but were then rapidly replaced by round and triangle hollow-out bands placed at intervals. In addition to being applied to decorate the stems of *dou*-stemmed dishes, these decorations also started to appear on small pottery objects such as the plates of *dou*-stemmed dishes, cups, bottles, and vases. They become the typical decorative pattern of the Songze Culture ceramics and are commonly found in the Greater Songze region. The carved patterns are usually decorated on the shoulders and waists of large pottery vessels such as pots and urns. Most scholars classify the carved patterns as leaf vein, broke-line, water ripple, fish-scale patterns, and so forth. These carved patterns usually have two or three parallel lines that coiled and spread over the surface of the object. The connecting part of the lines never crossover each other, but always overlapped, which resembles weaved bamboo splits, rattans, and willow twigs (Figure 5.1, nos. 1 to 4 and 14). Thus, Dr. Sheng, sensibly, I think, called these patterns, collectively, weaving patterns. Other carved patterns appear on the bottoms and lids of objects or on spinning wheels, which are usually regarded as symbols or family crests – such as the explanation of the famous octagon (*bajiaoxing*) pattern.[2] This kind of pattern has the same principle for line carving and coiling as the patterns found on the belly of some objects. The patterns which encircle the centre of these the object appear to be different. Thus, this kind of carved pattern is still a weaving pattern which imitates the bottom of baskets (Figure 5.1, nos. 11 to 13). The Songze Culture also has some painted pottery with similar weaving patterns. If we look at the engraved patterns again, it seems that it is also a kind of weaving pattern. The encased roundness and triangle holes could be regarded as the gaps found on braiding work.[3] The pottery *dou* high-stemmed dish found at the Zhili site, Anji County, strongly supports the preceding conjecture as it used for both engraved and carved weaving pattern for decorations (Figure 5.1, no. 3).

The preceding patterns and shapes, which could represent the ideas of the potters, are the dominant images that can be seen in the Songze-period pottery. Although sometimes slightly humorous, they are obviously intending to be depictions of daily life. There are also a few cups and pots with simple carved lines. The artists appear to be attempting to capture fishing and hunting scenes. However, these objects never present a serious conception as the workmanship is crude and the carved lines are hasty.

Jade is a rare resource that possesses an innate natural beauty. Thus, jade usually plays a special role in people's life. Songze is a society with jade worship tradition. More than a thousand jade objects (sets) have been recovered during the formal excavations in different areas. The majority of the jade objects are accessories such as *huang* semi-circular ornaments, *huan* rings, *jue* slit rings, beads, and *dang* pendants. *Yue* axe engraved with a hole likely originated from ordinary axes. During the Songze period, jade *bi* disk is small and probably used as an accessory as well. Thus, calling it *bi* here could be misleading. In addition, a number of *yue* axes, axes, and *ben* adzes have also been found.

A large number of these accessories were mainly unearthed from large tombs. Thus, these accessories not only embodied the aesthetic pursuit of the time but also presented the wealth status of the owner. *Yue* axes have only been found in large tombs and sometime placed with a couple to a dozen stone and jade axes and with *ben* adzes. During the Songze period, some axes became flat and thin. This might suggest their functional shifted from being used as tools to becoming used as weapons. A *yue* axe buried with an axe and *ben* adze means that the owner had had certain positions in both the military and production domains. The quality and quantity of the burial objects seems to embody the stratum of the owner in these domains.

As accessories and weapons, the jade objects reveal the background information of the Songze social structure. In addition, the jade objects with vividly carved symbols, such as human-shaped pattern, turtle-shaped pattern, bird-shaped pattern, beast-shaped pattern, and carved jade tablets, could help us better understand the mindset of the Songze people. Unfortunately, the number of such objects is very limited. Only approximately 40 pieces have been found at Lingjiatan in Anhui Province, where the highest-quality jade objects are present. Together, along with other scattered finds, the percentage of such jade objects in total amounts to less than 6 per cent of the whole assemblage.

Jade turtles have been found in two tombs at Lingjiatan (Anhui 2006a, 2008). The turtle is commonly found in burials in the Jianghuai region. Several tortoise shells with gravel inside have been recovered from the middle Neolithic cemetery of Jiahu (Wuyang, Henan Province). These tortoise shells seem to be related to the ancient *Shifa* divination practice and may be the oldest archaeological record of *Shifa* in China (Henan 1999). In 2007, one jade turtle and two turtle-shaped oblate objects were unearthed from tomb no. M20 at Lingjiatan. One to two jade sticks were found inside these objects. These burial objects should have the same function as the tortoise shells found at Jiahu.[4] Thus, we may speculate that these were primitive methods of prediction and related ceremonies. In addition, the owner would also have *yue* axe meaning that they possessed a dual identity of being both a warrior and a shaman. A jade pendant with carved symbols has been found inside another jade turtle at Lingjiatan. The decoration on the pendant was developed from the anise-shaped pattern (Figure 5.2, no. 4). Many scholars have explored the function of this jade turtle. Most arguments are centred around astronomical observations, observing orientation, and so on.[5] Other evidence for the anise-shaped pattern on jade objects comes from an object that looked like a combination between an eagle and a beast (Figure 5.2, no. 7). The anise-shaped pattern used on jade objects was originally decorated on the bottom of ceramics which imitated weaving products. The shape of the jade objects is irrelevant to the shape of daily-use pottery containers. From this phenomenon, we could know that the meaning of the anise-shaped pattern was transformed during this period. This pattern became a symbol but its function and meaning need further research. Furthermore, the *Huang* semi-circular ornament with two carved animal horns was still being used as an accessory. Only three *huan* rings (or *jue* slit rings) with dragon patterns have been found (Figure 5.2, no. 11). Those objects are small accessories.

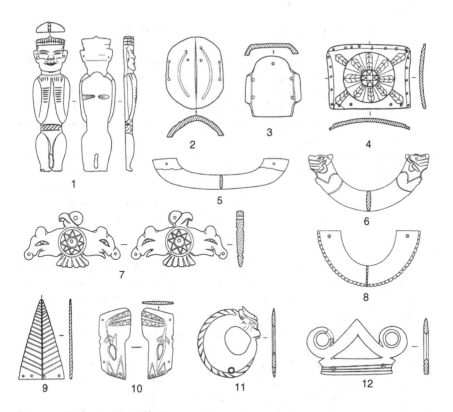

Figure 5.2 The Songze Culture jade objects with carved ornamentations found at the Lingjiatan site – 1. 98M29:16; 2. 87M4:35; 3. 87M4:29; 4. 87M4:30; 5. 87M4:80; 6. 87M8:26; 7. 98M29:6; 8. 97M8:27; 9. 87M4:68-1; 10. 87M4:40; 11. 98M16:2; 12. 87M15:38

Six jade figures were discovered from two large tombs at Lingjiatan. The front side of these figures was carved with representational portraits while the reverse of the figures had no decorations. A pair of connected holes might be used to slot the jade together with textiles (Figure 5.2, no. 1). Some academics believe that these figures represent the portraits of their worship (Zhang, J. 2006). However, other scholars have opposed this opinion by arguing that these figures are the portraits of the people who performed the worshipping (Fang 2006). According to the usage of these figures, it seems that the latter argument is the more plausible one. However, this argument needs more supportive evidence. Four human figures with *guan* coronets (or *chantou* decorating brocade around the head) and square faces have been discovered dating to the early Liangzhu period. The four Liangzhu figures are all small pendants with engraved patterns. Instead of being in sitting or standing positions, the postures of these figures involve what might be termed 'the delights of life' and possibly present some kind of storyline. It is also possible that these figures also show people's attires and postures at this time.

According to the preceding analyses, several observations can be made:

1   The majority of imagery materials dating to the Songze period are from ceramics. The decorations on the ceramics and the shapes of them are derived from daily matters and everyday life with humorous characteristics. This is the distinctive 'Songze Style'. This style presents that the life of common people, as least in the ideological domain, was not severely restricted.

2   According to the scale and hierarchy of settlements, as well as the discoveries at large settlements such as Lingjiatan (c.160 ha in size) and Dongshancun, Songze was a hierarchical society (Nanjing Museum et al. 2010, 2016). The status of the elites was mainly embodied in the possession of valuable objects. There are around 1,500 burials that have been excavated at the Lingjiatan site. A very clear hierarchical system in terms of quantities of burial goods can be seen among these burials. Of the five most extravagant tombs, tomb no. 07M23 yields more than 300 burial objects, more than 200 of which were jade items, alongside ceramic vessels, turquoise, and other objects (Anhui 2006a, 2008). Another example is the Dongshancun cemetery where several burials also contain larger numbers of burial goods while some other ones have much less objects (Nanjing Museum et al. 2016). The many decorative ornaments buried in their tombs represent their wealthy and their control of rare resources. The axes and *ben* adzes discovered in burials display their secular privilege. However, no specific objects related to worshipping could be discovered among the imagery materials of the abundant jade objects owned by the upper class.

3   If it is more appropriate to interpret jade turtles and other materials as tools for the *Shi* divination; thus, we can say that the Songze people

already had their own prediction system and associated ceremonies. This system is not only a primitive *fangshu* (arts of necromancy) but it also represents the Songze people's understanding of the cosmic order. For them, there should be a dominator which has the power to determine the result of the prediction – this could be a human, an object, or an abstract force. These may be the main contents of the social religious system in Songze society. At present, we do not understand this system in detail. As for the anise pattern found on jade objects, it is hard to regard it as a worship object. It seems to be a concept extracted from daily life, more likely to be an understanding of the Songze people's cosmic order. However, the real meaning of this system could not be understood at present.

In summary, the limited archaeological evidence on religions and beliefs mainly reflect the humility and ingratiation the Songze society showed to authorities. Although this phenomenon may be common among all religions, as the lack of detailed images of the worship object, we could only conjecture that the religion in Songze is primitive as it was close to natural worship. The Songze style discussed here presents activities of daily life under such a religious ideology.

5.3 Liangzhu mode

The Liangzhu Culture developed from the Songze Culture. No evident gap could be seen from the typological sequences between the two pottery assemblages. Thus, a 'transitional cultural period' has been designated to refer to the transition between the two archaeological cultures (Song 2000). However, the stylistic characteristics and assemblages of the pottery were changing rapidly after entering the early Liangzhu period.

During the early period, the weaving pattern as well as those *Xieshi* (representational)- or *Xieyi* (impressionistic)-style decorations already disappeared on the ceramics. In other words, the 'Songze Style' did not exist anymore. Except for the simple oblique carved lines on the 'T'-shaped tripod feet, the majority of ceramics during this period did not have any decorations. The plain ceramics provide a glimpse to the dull and hidebound lifestyle of the grass roots. Nevertheless, pottery production became more developed. The black pottery with a lustre shining surface became a fashion of the day. During the late period, some carved patterns, mainly birds patterns and dragon patterns, suddenly appeared. These patterns usually overspread the whole surface of the pottery. According to the carving technology, their appearances, and contents, these patterns were not developed from the typical Liangzhu-style ceramics. Rather, they were transformed from the decorations on jade objects (Figure 5.3). The polished surface of many Liangzhu ceramics creates a shiny, lustre effect very similar to the Liangzhu jade. Associated with this similarity are the shared motifs or decorative themes between the pottery and the jade.

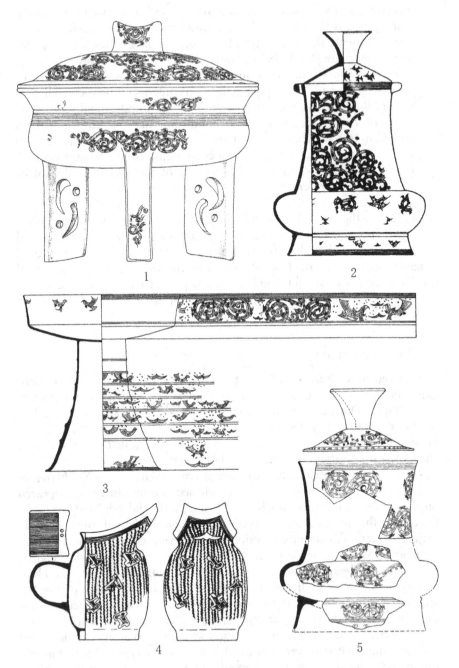

Figure 5.3 The Liangzhu pottery with carved ornamentations excavated from the Fuquanshan site – 1. M65:90; 2. M74:166; 3. M101:90; 4. M65:2; 5. M65:1

These suggest the Liangzhu ceramic industry received strong influence from the jade-making industry.

The Liangzhu Culture has pushed the custom of jade worship that they inherited from the Songze period to the peak. To date, the number of Liangzhu jade objects discovered by formal archaeological excavations has overwhelmingly outnumbered those found in the Songze Culture. Just a single burial (no. 20) at the Fangshan cemetery already yielded more than 500 jade objects. Liangzhu fully adopted jade manufacture technologies from the Songze, such as string cutting, tubular drilling, and cone drill. Liangzhu inherited the ensuing technology, a combination of string cutting and hollowing-out techniques, from Songze and advanced it during this time. In addition, Liangzhu people invented technologies to create relief patterns and for microscopic carving. Liangzhu inherited the basic types of jade objects from the Songze Culture, such as *huang* semi-circular ornaments, *huan* rings, *jue* slit rings, bracelets, *guanzhu* tubular beads and tablets, and so forth; the evolutionary paths of shape from the Songze jade objects to the Liangzhu jade objects is quite clear. The Liangzhu Culture also created some new types of jade. For example, Liangzhu people developed small *bi* disks in Songze into large *bi* disks of their own style. They also changed the Songze bracelets and created a new shape: *cong* tubes. As both of the shapes of these new products in Liangzhu have clear differences from their prototypes in Songze and they all lost its function as decorative ornaments during the Liangzhu period, it could be better to regard them as new types of jade objects. Other types, such as *zhuixingqi* awl-shaped objects, *sanchaxingqi* three-pronged object, *guanzhuangshi* cockscomb-shaped objects, and *zhuxingqi* cylindrical objects are the new creations from the Liangzhu (Figure 5.4). There are more than 40 types of jade objects in Liangzhu in total, which is far more abundant than Songze jade objects. A more significant change of the Liangzhu jade objects is the prevalence of carved symbols. Although jade figures, jade dragons, and jade eagles (birds) could also be found in the Songze Culture, their shapes are all in a representational style, which is quite different from their Liangzhu counterparts. From the perspective of archaeological typology, several developmental phases are missing to be able to link the jade objects of the two cultural periods smoothly. Thus, we could only regard the Songze Culture as the ideological origin of the shapes and decorations of the Liangzhu jade objects. The most striking decoration in Liangzhu is the animal-face motif as well as the sacred human and animal-face motif. Currently, the origins of these decorations could not be found in the Songze Culture. Fang points out that the features of Liangzhu jade figures are similar to the Songze jade figures; the animal-face motif may be related to the tiger-shaped motif from Songze (Fang 2006). However, the Songze animal-face images reconstructions are still significantly different from the one found in Liangzhu, let alone the Liangzhu's idea of decorating the animal face on the most conspicuous parts of the objects. Thus, these face-shaped motifs could also be regarded as the creation of the Liangzhu people.

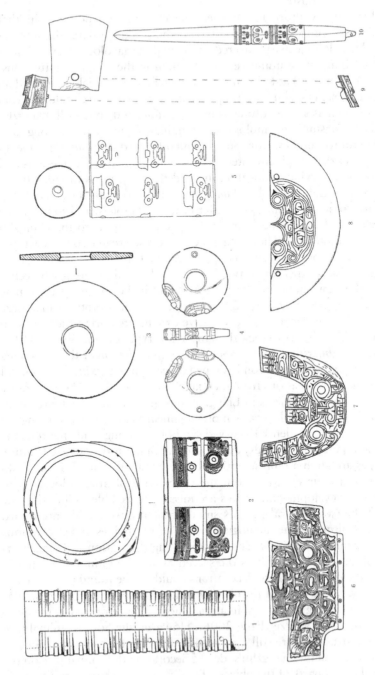

Figure 5.4 The types of the Liangzhu Culture jade objects and their ornamentations – 1. Sidun M1:6; 2. Yaoshan M2:22; 3. Sidun M1:1; 4. Fanshan M22:26-1; 5. Fanshan M12:87; 6. Fanshan M16:4; 7. Yaoshan M3:3; 8. Yaoshan M4:34; 9. Yaoshan M7; 10. Fanshan M20:73

According to the far-flung and profound changes of the aforementioned ceramic and jade objects, the society seemed to have experienced dramatic organisation since the Songze period. Evidence from recent archaeological surveys (Anhui 2006a, 2008) show that the Lingjiatan site, a site of the highest level in the Greater Songze region, was abandoned during the late Songze period. Immediately after this was the establishment of several sites at nowadays Liangzhu and Pingyao towns, located to the south of the Tianmu Mountain, and north-west Hangzhou during the Liangzhu period. Later in the Liangzhu period, a large site cluster appeared in the same area. An enormous urban centre and a large hydraulic system outside the centre were established (Zhejiang 2015b). The centre for jade manufacture was also transferred to here. The 1,400-m long city wall, the large artificial platform at Mojiaoshan (600 × 400 m), and the hydraulic system to south side of Tianmu Mountain (about 6,000 to 7,000 m long in total) were not only unprecedented gigantic infrastructure projects but also clever engineering designs. The society presented its ability in mobilising and providing technologies, labour input, and other resources as well as its organisational and administrative capabilities.

Along with the profound and dramatic change of the Liangzhu society, the Songze Style has now been replaced by a decorative system using sacred human and animal face, bird, and dragon as the main themes. This phenomenon signifies the new ideology of the Liangzhu people. When experiencing significant transitions, human society came out with new ideologies, religions, and beliefs to better fit into new conditions. These phenomena could be found in times when the Israelis established their early country, when the Aryans expanded to the subcontinent, and during the spring and autumn and Warring States periods of political turmoil and upheaval (Armstrong 2010). The transition from Songze to Liangzhu provides another useful case for this point.

What are the actual contents and characteristics of the Liangzhu ideology? Let us come back to its rich image materials.

There is large variety of jade objects with carved motifs in Liangzhu. However, the themes of those patterns are surprisingly limited and only have portrait (god), spiritual beast, bird, and dragon. As for other patterns, such as the swirl pattern, they serve as the background for the expression of the thematic motifs just mentioned. We could not treat them the same as those thematic motifs. Among the limited thematic motifs, the sacred human and animal-face motifs are the most common ones. Usually, these motifs were used separately from other motifs. Even when combined with the bird patterns, they are still the main subject on the jade objects. Their significant symbolic status suggests that they were the most important worship objects of the Liangzhu people. By the end of the middle Liangzhu period, bird patterns were regarded as the subordinate motifs decorated by the side of the sacred human and animal-face motifs. In the burial objects found at the Fanshan cemetery, there is a kind of bird-shaped pendants. Different from the refined and elegant objects such as *cong* tubes and *yue* axes, similar pendants of turtle shape and fish shape were all used as decorative ornaments. Obviously, in Liangzhu people's mind, the sacred human and animal-face motifs are more important than the bird motifs. Once again, the dragon patterns have less significance than bird patterns as they never appeared on valuable objects such as *cong* tubes.

Multifaceted examination of the Liangzhu jade motifs, including their types, the frequencies, and the relations between the decorations and object types as well as the combinations of different decorative themes, clearly suggests that the sacred and animal-face motifs are the main subject of Liangzhu belief system. It might not be the only subject, but it must be the most important worship object. Thus, the uniqueness of the worship object in the Liangzhu religious system is closer to the 'monotheism' in the history of religion (Goodman 1981).

What is the content of the Liangzhu religion? How is it formed?

If we look at the sacred and animal-face motif carefully, we could find that it consists of two subsystems. The first one includes single human-face motifs and single animal motifs. Each has its own complete contour and could be used individually, such as the animal-face patterns on pendants. Those single-sectioned *cong* tubes that have human-face motifs or animal-face motifs are the examples from the earliest period. For example, the *congshizhuo cong*-style bracelet with an animal face found at the Zhanglingshan site (Figure 5.5, no. 2) could be dated to the earliest period of the Liangzhu Culture or even to the 'transitional period' between Songze and Liangzhu based on the typologies of the motifs (see Chapter 3). During this period, the classic *cong* tubes had not appeared yet. Thus, this object is the prototype of later period *cong* tubes and the animal-face decorations as well as the origin of the idea of combining object shapes and decorations. Based on the stylistic and typological changes of motifs, the object no. M7:50 from the Yaoshan cemetery is the earliest representation for the *cong* tube with a human face decoration (Figure 5.5, no. 3). This object seemed to appear slightly later than the *congshizhuo cong*-style bracelet found at Zhanglingshan. But their dates are very close. In addition, except for *cong* tubes (including small ones), human-face motifs seemed to have never appeared on other types of objects. This type of motifs could be the dedicated decoration for *cong* tubes.

Cong tubes were the most incomprehensible objects created by Liangzhu people. Liangzhu people also created two decorative motifs of great significance for this type of objects. It is difficult to assess which motif is more important in the early period as both motifs could be used to decorate objects individually. We could only conjecture that the human face is the personification of Liangzhu people and the animal face may represent the mysterious power from the nature. However, along with the appearance of the multisectioned *cong* tubes, these two motifs, contrastingly different in terms of their original meanings and artistic designs started to appear on one object. Normally, the human face was carved on the top of the animal

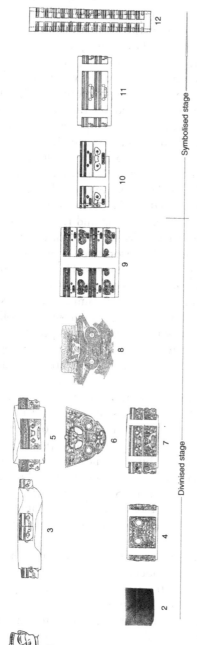

Figure 5.5 The formation and evolution process of the 'Liangzhu Mode' – 1. Liangjiatan 98M29:14; 2. Zhanglingshan M4:02; 3. Yaoshan M7:50; 4. Yaoshan M9:4; 5. Yaoshan M10:16; 6. Yaoshan M10:20; 7. Yaoshan M10:19; 8. Fanshan M12:98; 9. Fanshan M12:93; 10. Fuquanshan M65:50; 11. Fanshan M23:22; 12. Sidun M3:26

face (Figure 5.5, nos. 7, 9, and 10). The implication behind this phenomenon could be understood by a closer examination of the sacred and animal-face motif, a combination of the human face and animal face.

The sacred human and animal-face motif appeared at the same time with the multisectioned *cong* tubes, slightly later than the appearance of the single-sectioned *cong* tubes (Figure 5.5, nos. 6 and 8). It was widely used and could be found on the *cong* tubes, *yue* axes, and *zhuxingqi* cylindrical objects as well as the ivory sceptre found in the large tomb at the Fuquanshan site. It was also applied on the *guanzhuangshi* cockscomb-shaped objects and the *sanchaxingqi* three-pronged objects. But the number of these objects is very limited. This motif usually appears as a figure, arms akimbo, with finery and a tall and gorgeous feather crown. Obviously, the Liangzhu people made every effort to beautify the figure. The lower part of the figure is a squat beast. The posture of the beast is fierce but tame. In other words, the beast is reined by the person seated on it. Intuitively, this image tells the story of a human taming and reining a beast. However, the Liangzhu decorative system has a distinctive feature of abstraction. The god, beast faces, dragons, or birds the decorative system tried to present are not realistic facsimiles to their original images. Thus, the preceding pattern is more likely to be an abstractive concept or idea which is sublimated based on the story. Liangzhu people was very familiar with this story as it must have been retold over and over again through this symbolised motif, which mainly appeared on the *cong* tubes. If Liangzhu people had made creations on jade decorations, they were mainly about the carving techniques. No aspirations could be found in exploring and creating new decorative systems by the Liangzhu people other than sticking to the sacred human and animal-face motif. Thus, we could regard the abstractive sacred human and animal-face motif as the idea of 'human reins the nature'. It is not coincident that this kind of motif appeared when the Liangzhu people were establishing the brand-new society after the Songze period. This pattern seemed to eulogise the heroes in the fights to establish the new cosmic order and express the heroic feelings of 'human will concur the nature' (Luo 2001). Such confidence could not be found on the Songze people who used to obey and cater to the cosmic order as we have just discussed in the previous section. The collective consciousness was completely changed after entering into the Liangzhu period.

The ideology of Liangzhu people was rapidly transformed during the early Liangzhu period. One such evidence of this is the new imagery representation system that appeared in the archaeological record. This system includes special jade objects such as *cong* tubes and *bi* disks. As mentioned, this system appeared so fast yet was already complete, implying that it might have been a system specially designed by somebody on the top of the society, rather than an outcome of long-term evolution. If the latter was the case, there should be a series of materials in the archaeological records which could present the transitional process from representational and realistic to mysterious and abstract styles of objects, such as the continuous imagery evolution of the Yangshao Culture, from the representational fish patterns to pictorial fish patterns and then to geometric patterns. However, such material evidence could not be found in the motif system of Liangzhu so far.

The subsequent changes further explain how this new ideology was consolidated and strengthened after its creation. The evolution of the Liangzhu jade decorations experienced two main stages. The first stage is from the early Liangzhu period, presented by the material from the Zhanglingshan and Dongshancun sites, to the middle Liangzhu period, represented by the archaeological assemblages at Yaoshan, Fanshan (tomb no. M12), and Fuquanshan. During this stage, the image of the sacred human with animal-face motif was transformed from a rough to a delicate style involving micro carving and related carving techniques. The advancement of this decorative system was accompanied with the development of jade production technologies. The main purpose of this development might be to mythologise or deify the protagonist of the story. The sacred human and animal-face motif disappeared from *guanzhuangshi* cockscomb-shaped objects, pendants, and other daily-use objects after this stage. Instead, it became the exclusive decoration for important objects such as *cong* tubes, through which it then gained pure godhood. By the late Liangzhu period, the style of the sacred human and animal-face motif was once again changed and became more and more succinct and abstractive. The motif finally lost all detail designs, including animal tusks, and became a completely simplified symbol (Figure 5.5, nos. 11 and 12). This process may imply that the story behind the motif was minimised, even though it continued to linger upon the collective consciousness as a concept understood by the Liangzhu people.

A further issue is, why did Liangzhu people create this system and carefully maintain it?

Another striking character of the Liangzhu imagery material is its universality and consistency. *Cong* tubes and other valuable objects are commonly found in the distributional areas of the Liangzhu Culture across the Yangtze Delta. This phenomenon is quite different from the Songze period when the special jade objects could only be found at Lingjiatan. The Liangzhu jade objects from different places are highly consistent in shape, decorative theme, and the ways how the motifs were manifested. Given the scarcity of raw materials and the complexity of the jade-production technologies, many researchers believe that the Liangzhu jade objects are manufactured by one or at most only very few production centres. Jade objects around the region are highly consistent in shape and motifs. There were fixed collocations between the decorative themes, the type of objects, and the place where the motifs should to be decorated. This consistency suggests that the production system followed strict regulations. The Liangzhu craftsmen may have the talent for artistic creations like Songze people, but they did not have the permission or aspiration to do so. Most researchers agree that the manufacture of the Liangzhu jade objects was controlled by members of the upper class. Thus, same with modern industrial prints, those duplicated jade objects and decorations express the will of the upper class. In other words, the elite group drove the creation of the Liangzhu ideology. The widespread of the Liangzhu jade objects over the region once again proved the wide acknowledgement of this ideology, at least among the upper classes of different regions. This propaganda and education method invented by the Liangzhu elites unified the social consciousness of the Liangzhu society.

I once noticed that the Liangzhu society had a strong religious atmosphere. Religion was not only used for unifying social consciousness in Liangzhu. In the highest-class tombs of Liangzhu, the tomb owners were found to own numerous objects such as *yue* axes and stone *yue* axes as well *cong* tubes and *bi* disks. These special objects have large volumes (meaning they used a great deal of raw material) and their functions could only be explained from a religious perspective. The *yue* axe found in Fanshan tomb no. M12 has the sacred human and animal-face motif on it, which is rare as such motifs are more commonly found on *cong* tubes. These phenomena all indicate that the religious privilege was closely integrated with the power in military and social management. In the real life, the Liangzhu elites had a dual identity by owning the secular power and the priesthood (Zhao 1999). Nakamura further argued that the highly consistent jade objects found in the Liangzhu region were manufactured by the craftsmen from the Liangzhu City. By redistributing those objects to local authorities, the elite group at the Liangzhu centre recognised the regional dominions and obtained the supports and acknowledgements for their central leadership in return (Nakamura 2003). Thus, the original idea of the Liangzhu religion was to realise the control over the society. It is a religion with political overtones or a 'political religion'.

To sum up what I have analysed so far:

1　According to the typological analysis on the imagery materials, a decorative system appeared during a short period immediately after the appearance of the Liangzhu Culture. This system was completely different from the Songze one.
2　This decorative system conveyed Liangzhu people's delight from the success in establishing a new social order and their glories for human power. They embodied this sentiment in a heroic image of reining beasts and worshiped the image.
3　The prototype of the hero was very likely to be the ancestor of a blood lineage. He was deified and finally became the most important or even the only worship object. Since then, the religion gained its doctrine of worshiping one God or developed into a monotheism.
4　This social consciousness was the design from the upper class and gained the acknowledgement across the whole society (probably by the means of assigning jade objects and other ways). It finally reached its original purpose of conforming the society.

5   The religion in Liangzhu played an extremely important role in the social life, deeply involved in the management of all social levels and domains. Thus, in terms of its purpose and function, the essence of the Liangzhu religion is a 'political religion'.

According to the man-made design and management behind the characteristics of the Liangzhu decorative system, we could summarise these characteristics as the 'Liangzhu Mode' to be distinguished from the 'Songze Style' which mainly used nature expressions.

5.4 Conclusions

In terms of cultural traditions and the blood lineage, the Songze Culture and the Liangzhu Culture are the material remains from the people and their generations who lived in the Lower Yangtze River region during these two continuous periods. In this sense, they are an integrated entity. However, both the 'Songze Style' and the 'Liangzhu Mode' have distinctive characteristics. Together with the transition from the 'Songze Style' to the 'Liangzhu Mode', their appearances indicate that they were rooted in different social conditions. Drastic changes were taken place throughout their history. It might be better to study their connotations by comparing the creation of the faith of Judaism.

Since around thirteenth century BC, the ancient Israelis, scattering in Canaan, started the course of integrating each clan and establishing a unified country. The process lasted more than two centuries. During this process, the status of Jehovah was highlighted from the original theocracy system and became the uppermost faith of the ancient Israelis. In 1030 BC, Israelis finally established their united kingdom. The historical philosopher Eric Voegelin regards the gradually highlighting of Jehovah's status as a method for the societal development expressing its aspirations to the new order; in addition, the establishment of the new order was promoted and accomplished by believing and worshiping Jehovah (Voegelin 2010). This is the dialectical relationship between ideology and social development. Based on the same idea and the analysis on the imagery materials from archaeological remains, we have summarised two different ideologies, the 'Songze Style' and the 'Liangzhu Mode'. Through these we could touch upon the essential contents of the social change from Songze to Liangzhu and acquire a new understanding as well as explanation for prehistoric Lower Yangtze River. This is the real purpose of this chapter.

However, restricted by the space, another important topic has not been discussed in this chapter, which is to contextualise the Songze and Liangzhu cultures in a broader historic context. During these two cultural periods, contemporary societies in other parts of China also accelerated their processes to social complexity. The diversity of the appearances of those material cultures strongly implies the diversity of the societies (Liu and Chen 2013) and the reasons, mechanisms, methods, and strategies of their civilisational discourse. If we compared Songze and Liangzhu in this wider context, we could better understand the characteristics of the prehistoric societies in the Lower Yangtze River. Meanwhile, we could also obtain understandings on the histories of other societies, which will be discussed in my articles elsewhere.

There is another question that has not been touched upon yet. This chapter emphasises the transition from Songze to Liangzhu. However, according to the discovered archaeological materials, since the formation of the 'Liangzhu Mode', although it was carefully maintained by the Liangzhu people for a long time, changes did happen inside this model, such as the transitional process of the sacred human and animal-face motif from the exaggerated deification to the symbolism and conceptualisation. If we regard it as a continuous process, then the changes happened in the late Liangzhu period become more apparent and worth noticing. For example, fine ceramics covered with carved dragon motifs or bird motifs and the new carved motifs for *bi* disks, the 'bird standing on a platform', appeared during the late Liangzhu period. Meanwhile, the significance of the sacred human and animal-face motif seemed to be weakened. These phenomena are related to another important question: the disintegration of the Liangzhu society and the historical contributions of Liangzhu to the emergence of civilisations in prehistoric China. With the accumulation of more and more archaeological materials, these questions may be answered in the near future.

Notes

1 Many researchers agree with this opinion. However, no supportive work has been published on this topic to date.
2 There are a number of publications about the octagon patterns, especially the one on the jade painted block found at the Lingjiatan site. The most significant works are published in the *Lingjiatan Wenhua Yanjiu (Research on Liangjiatan Culture)*.
3 When I was a student, I attend a lecture given by Zhengduo Wang. He argued that the exaggerated hollow on the handle of *dou*-stemmed dishes from the Dawenkou period is an imitation of bamboo-weaving products. Also, '*bian*', the alternative name of *dou*-stemmed dishes during the pre-Qin period, also hints at the relationship between the shape and ornamentation of pottery *dou*-stemmed dishes and bamboo-weaving products. Same arguments could also be found (Liu 2012). Liu's argument is quite valuable and sensible as the Dawenkou Culture Huating period appeared after the Songze Culture and thus it could be seen as the influence of the Songze Culture in northern China.
4 In ancient China, Shi and turtles were always mentioned together. However, *The Book of Rites (The Quli Volume)* (Dai 2008) said that 'turtle is used for *Bu* divination and *Jia* is used for *Sh* divination'. '*Jia*', or '*Ce*', is the alternative name for the grass used in divination, called '*Shi cao*'; '*Shi* (筮)' is the old name of '*Shi* (蓍)', here it means a method of divination. Based on these records we know that '*Shi*' and '*Bu*' are two different methods. In ancient Chinese classical texts, all turtles

were related to divination. Turtle is a tool for divination. The oracle turtle bones dating from the Shang dynasty are good evidence supporting this idea. However, at Jiahu and Lingjiatan, turtles are the containers for gravel and jade sticks. Turtles are also related with the 'Shi' divination practice. This phenomenon could not be found in the textual evidence. A number of oracle animal bones have been found in north China that could be dated back to the late Neolithic period. However, no turtle oracle bones have been found so far. In south China, no animal oracle bones or burned turtle oracle bones have been found. Thus, we could conjecture that using turtle oracle bones for '*Bu*' divination practices was developed later than late Neolithic periods. The opinion that 'the people in ancient south China practiced *Shi* divination while the people in north China practiced *Bu* divination' might be true (Chen 1996, pp. 78–82).

5 See note 2.

6 Shamanistic, historic, and virtuous jade

Continuity and change in early Chinese jade traditions

Sun Qingwei, translation by Catherine Xinxin Yu

6.1 Introduction

When discussing the rituals of the 'Three Dynasties' (the Xia,[1] Shang, and Zhou dynasties), Confucius said,

> The Yin (Shang) Dynasty followed the rituals of the Xia; that which was abolished or added may be known. The Zhou Dynasty followed the rituals of the Yin; that which was abolished or added may be known. Some other may follow the Zhou, but though it may be at the distance of a hundred ages, its affairs may be known.
>
> *The Confucius Analects: Wei Zheng* (Yang 2006)[2]

Although ostensibly Confucius seemed to be emphasising the changes to the ritual system throughout the Three Dynasties, he was actually trying to see past these changes and grasp the factors that remained constant as history unfolded.

Traditional Chinese society revolved around rituals. In the article *Explaining 'li' (ritual)*, Wang Guowei stated that the original meaning of the character *li* was 'a vessel containing jade to serve gods', therefore 'the ancient Chinese used jade for rituals' (Wang 1959).[3] By comparing this theory to archaeological evidence, one would discover that early Chinese jade was not only an integral part of the ritual system but it also underwent many changes. Three traditions were gradually formed: *Wu* (shamanistic) jade, *shi* historic jade, and *de* virtuous jade. I will attempt to provide an analysis of these three traditions in the following text.

6.2 Prehistoric period: The rise of the *wu* (shamanistic) jade tradition

Generally speaking, the ancient Chinese started using jade from the Neolithic period onwards. Jade objects appeared in the early Neolithic, developed during the middle Neolithic, and reached their first peak during the late Neolithic.[4]

According to archaeological evidence, multiple centres of jade-use emerged during this first peak. Deng Shuping summarised them as having three origins of ancient Chinese jade cultures: the 'Eastern Area' represented by the Hongshan Culture in northeast China; Dawenkou Culture and Longshan Culture in Shandong; the 'Southeastern Area' represented by the Liangzhu Culture (Figure 1.2); and 'Western China' located around the Upper and Middle Yellow River area (Deng 1995, 2007). However, as for having a clear and well-established origins supported by archaeological evidence, only two jade traditions during this period would qualify: the Hongshan jade culture in the north and the Liangzhu jade culture in the south. The former was a successor of the Chahai and Xinglongwa jade cultures (Shelach and Teng 2013), while the latter was closely connected to the use of jade in the Songze, Majiabang, and Lingjiatan cultures. In comparison, the jade tradition in Western China developed slightly later, and no known site in this area is comparable to Niuheliang of the Hongshan Culture or Fanshan of the Liangzhu Culture with the most elaborate use of jade. Therefore, some scholars believe prehistoric Chinese jade should be divided into two areas instead: Liangzhu in the south and Hongshan in the north (Huang 1987).

For a long time, small amounts of jade objects were frequently excavated within the area occupied by the Hongshan Culture. During the late 1970s and early 1980s, large-scale Hongshan sites were found at Niuheliang and Dongshanzui in western Liaoning Province. There are remains of ritual structures such as 'goddess temples' and burials in stone altars (*jishizhong*). Exquisite jade was the only kind of funerary object placed in the large central pits within these burials, an archaeological phenomenon called 'burial with jade only' by some scholars. Common object types include jade *yuzhulong* pig-dragons, *gouyunxingqi* hook-and-cloud ornaments, *matixingqi* horse hoof-shaped objects, comb-shaped ornaments with animal-face motif, *bi* disks, and *huan* rings (Guo 1997; Liaoning 2012). Doubtlessly, the occupants of these large tombs in the centre of the stone platform burials belonged to the highest class in Hongshan society, as jade objects were symbols of their power and wealth and possibly also a unique tool for them to communicate with gods (Guo 1998; Nelson 2008; Peterson and Lu 2013).

Jade objects are also widely found in the area occupied by the Liangzhu Culture. Some elite tombs show a resemblance to the 'burial with jade only' practice in the Hongshan Culture. Some scholars refer this as 'jades for burial' in Liangzhu period (Wang 1984). The 'jades for burial' phenomenon is particularly prominent in key sites that have both altars and cemeteries, such as Fanshan, Yaoshan, and Fuquanshan. For example, jade objects comprise more than 90 per cent of burial goods at the Fanshan cemetery, also known as the 'royal cemetery' of the Liangzhu Culture, demonstrating the prestige of jade in Liangzhu society (Shanghai 2000; Zhejiang

2003, 2005). As for the shape of objects, ornaments such as *huang* semi-circular ornaments, bracelets, tubes, and beads were dominant during the early Liangzhu period, while from the middle Liangzhu period onwards, diverse and marvellous shapes abounded, such as ritual objects like *cong* tubes, *yue* axes, *bi* disks, and ornaments like *guanzhuangshi* cockscomb-shaped objects, *sanchaxingqi* three-pronged objects, *zhuixingqi* awl-shaped objects, *huang* semi-circular ornaments, tubes, bracelets, and belt hooks, as well as animal-shaped objects in the form of turtles, birds, fish, and cicadas. Compared to Hongshan jade objects that are characteristically with plain surface, Liangzhu ritual jade objects such as *cong* tubes, *yue* axes, and *bi* disks are particularly eye-catching as they are intricately carved with the 'sacred human and animal-face' motif, though unfortunately no unanimous interpretation of its meaning has yet been reached within academia (cf. Liu 2007).

As for Western China, although there was no prominent centre of jade use during this period comparable to the Niuheliang, Fanshan, and Yaoshan sites, jade objects already showed distinctive regional characteristics. Examples include ritual objects and ornaments like *cong* tubes, *bi* disks, *huang* semi-circular ornaments, and *lianbi* multi-piece disks, and tool or weapon-style jade like axes, *ben* adzes, chisels, and blades from the Caiyuan Culture in Ningxia (Ningxia and Historical Museum of China 2003) as well as the Majiayao Culture and the Zongri Culture in Gansu and Qinghai provinces (Yan 2010). These abundant archaeological finds are reliable proofs that there was a unique jade culture in Western China during the late Neolithic period (Deng 2007).

Judging by object shape and raw material, it is clear that these three jade traditions developed independently, but they also share a significant common trait, which is the clear separation between ritual and everyday jade. More specifically, *yuzhulong* pig-dragons of the Hongshan Culture; *cong* tubes and *yue* axes of the Liangzhu Culture; and *cong* tubes and *bi* disks in Western China had a prestigious status over other ordinary objects. The privilege of these ritual jade items is manifest by their close association with the elite groups or people who had access to valuable items as they are mostly recovered from elite tombs and/or altars or ceremonious features. This distinction is particularly prominent in the Liangzhu Culture because *cong* tubes and *yue* axes were not only dominant ritual objects, but they were also consistently associated with the prestigious 'sacred man and animal-face' motif. Judging from this, the Liangzhu jade culture can be considered the most advanced among the three contemporary jade traditions. Moreover, the Liangzhu walled site can be regarded as the physical representation of the ancient Liangzhu state, whereas there is no archaeological evidence to date to prove the existence of comparable ancient states at either Hongshan or western China during this period.

Although everyday jade objects of the three jade traditions obviously have different shapes, they are the same in essence because most of them were ornamental objects. In other words, jade objects were used to decorate the body. Jade originally referred to 'beautiful stones possessing five virtues' (Dai 2008), as recorded in the *Book of Rites*, a Confucius cannon written between the end of the Warring State and Early Western Han period. Mankind spent a long time in the Stone Age, a period during which people were prone to worship stone and perhaps had the deepest understanding about the characteristics of various types of stone. As such, becoming devoted to a beautiful stone and using it to decorate oneself was almost instinctive to mankind (Zhang 2003).

The reason why ornamental and ritual objects were equally important during the prehistoric time can be attributed to the fact that ancient people extended their own preferences to gods. Sima Qian stated in *Records of the Grand Historian: Book of Rites*,

> I visited the 'Daxing' Official of Rituals to examine the rise and decline of rituals throughout the Three Dynasties, wherefore I understood that it was ancient practice to stipulate rituals according to common practice, and to conduct ceremonies according to human nature.
>
> (Sima 1982)

In other words, gods were served according to the likings of mankind ever since the days of the ancients. More specifically:

> Carriages please the body, therefore coaches are gilded and beams inlaid to elaborate the decoration. Colours please the eyes, therefore intricate colours and patterns decorate formal robes to represent people's capabilities. Instruments please the ears, therefore the eight instruments are sounded in harmony to move the heart. Tastes please the mouth, therefore bountiful delicacies with diverse tastes are concocted for enjoyment. Precious and refined objects please the soul, therefore gui tablets and bi disks are fashioned to convince the mind.
>
> *The Records of the Grand Historian: Book of Rituals* (Sima 1982)

The fundamental reason for conducting rituals and ceremonies was to serve the gods and pray for fortune. Sacrificial offering aimed at pleasing and entertaining the gods was the main method for achieving this goal. To please the gods, naturally the ancients had to offer to the gods the best things according to themselves, so as to satisfy all the sensory needs of the gods. Carriages, clothing, music, animal sacrifice, grains, alcoholic drinks, and precious objects all resulted from this idea. During the prehistoric time, the most beautiful and sacred thing was none other than jade. It is, therefore, inevitable that jade became one of the earliest kinds of ritual objects in prehistoric China such as the jade ornaments discovered in the burials of the Xinglongwa Culture of western Liao River 8,000–7,000 years ago (Shelach and Teng 2013).

Because religious activities were indispensable in early society, the role of a person specialised in religious matters, *wu* or shaman, emerged very early on. According to the explanation given by Guanshefu in *Discourses of the States: Discourse of Chu II*, the '*wu*' shaman is 'one who is spiritual and focused' (Xu 2002). 'The males are called *xi* and females called *wu*.' Their responsibilities were

> to decide on the location of the altars of gods and their order of import-
> ance, to prepare sacrificial animals, ritual objects and clothing appro-
> priate to the season, and to find he who is an accomplished descendant
> of a sage and knows the names of mountains and streams, the gods of
> ancestors, the affairs of ancestral temples, the order and arrangement
> of ancestors, and is respectfully diligent, proper in etiquette, prim in
> manners, respectable in appearance, honest and reliable in character,
> immaculate in dress, and is respectful to the gods, and make him the
> conductor of ceremonies.

Explaining and Analysing Characters: Jade Section defines *ling* (lit. 'spirit') as a *wu* shaman who 'serves gods with the use of jade' (Xu 1963). Hence, the *wu* shaman has two basic characteristics. First, he or she should be naturally gifted and is more spiritual than normal people, being 'one who is spiritual and focused'. Second, he or she should understand the means to communicate with gods and spirits, the most important among them being 'to serve gods with the use of jade'. Therefore, we might very well refer to prehistoric ritual jade as *wu* (shamanistic) jade in terms of their functions and symbolic meanings. If we examine jade pig-dragons from Hongshan and 'sacred man and animal-face' motifs on Liangzhu jade with this idea in mind, we would discover that they are steeped with pre-historic *wu* (shamanistic) rituality.

Kings acted as the head of *wu* shamans in some of the early stratified societies of China. In the so-called king's tomb at Taosi, for instance, not only were large number of exotic items such as lacquer objects and beau-tiful painted pottery were found, some objects that might be used in ritual activities such as stone musical instruments were also present. This suggests that the owner of this large and rich burial might be involved in some ritual performance. This phenomenon is also well recorded in literature. Famous examples include King Zhuanxu who 'stipulated right and wrong in accordance with spirits and gods' (*Book of Rites by the Older Dai: Virtue of the Five Kings*) (Dai 2005) and 'cut off direct communication between heaven and earth' (*Discourses of the States: Discourse of Chu II*) (Xu 2002) as well as 'many *wu* shamans walk with a limp like King Yu' (*Exemplary Sayings: Chong Li*) (Wang 1996) and King Tang who 'offered himself as sac-rifice to pray in the Sang Forest' (*Lv's Annals: Almanac of the Third Month of Autumn*) (Lv 1996). Li Zehou believes that the key to the intellectual history of ancient China is that, through the syncretism between *wu* shaman and king, between religion and politics, the idealised version of the basic characteristics of *wu* shamanism became the basis of traditional Chinese thought.

(Li 2012)

The use of jade in the Hongshan and Liangzhu cultures shows clear hierarchical differentiation. The most numerous and the finest jade, as well as some special shapes, have only been found in elite cemeteries such as Niuheliang and Fanshan. This is certainly the best indication of the syncretism between *wu* shaman and king at the time.[5]

6.3 Longshan period: Transition from *wu* (shamanistic) jade to historic jade

While the use of jade was dominated by three major traditions during the late Neolithic period, the onset of the Longshan period brought about a stark change to the situation. With the decline of Liangzhu and Hongshan cultures, distinctive jade cultures developed in the Qijia Culture, Taosi Culture, and Longshan Culture along the Yellow River, and the Shijiahe Culture in the Middle Yangtze River.

During the previous period, the jade industry from Western China was clearly less developed than that of the Hongshan and Liangzhu cultures. The raw material was less available, crafting technologies were less sophisticated, and there were less varieties of shapes of jade (Deng 2007; Yan 2010).[6] However, while the latter two fell into rapid decline at the beginning of the Longshan period, the former maintained continuous development, of which Qijia jade is a representative example. As Deng Shuping has pointed out, among all properly excavated Qijia culture sites, the Shizhaocun site in Gansu Province and the Lajia site in Qinghai Province have yielded the most interesting jade (Deng 2007, 2014). A *bi* disk and *cong* tube from tomb no. M8 at Shizhaocun were made from very similar materials, possibly the same piece of raw jade (Institute of Archaeology 1999a). Tomb no. M17 at the Lajia site is located on the altar at the centre of a square, a location that emphasises its unique status. Funerary goods in this tomb include nothing other than 15 jade objects, among which there are two *lianbi* three-piece disks and two *bi* disks (Institute of Archaeology Gansu-Qinghai Research Team 2004). Also, two *bi* disks were placed against the wall in house no. F4 at the Lajia site, equally pointing to the special significance of *bi* for residents of the Lajia settlement (Institute of Archaeology Gansu-Qinghai Research Team 2002). A set of four *cong* tubes and four *bi* disks in Qijia Culture style have been found in a sacrificial pit at Houliuhecun in Gansu Province (Deng 2014). Combined with similar objects in the collections of local museums in Gansu, Qinghai, and Ningxia provinces (Deng 2007, 2014; Cao 2014), they clearly demonstrate the existence of a jade tradition centred around *bi* disks and *cong* tubes.

Downstream along the Yellow River, the Longshan-period Taosi Culture in southern Shanxi Province also has a highly developed jade culture. More than 1,300 tombs were excavated at the Taosi site in the 1970s and 1980s, up to 200 of which contain jade funerary goods (Gao 1998). Based on current archaeological evidence, *yue* axe is the main ritual object of the Taosi Culture, while *bi* disks, *cong* tubes, *huang* semi-circular ornaments, and *lianbi* multi-piece disks are the major types of ornamental jade. According to Gao Wei, more than 80 *yue* axes have been excavated at the Taosi cemeteries in the 1970s and 1980s, usually only one piece per tomb, and mostly seen in the tombs of male occupants. More than 80 *bi* disks have been excavated as well, mostly placed on the chest and abdomen or near the arms and wrists of the tomb occupants, clearly as ornaments. Thirteen *cong* tubes excavated here were also placed near the tomb occupants' arms, only one piece per tomb and mostly seen in male tombs. To date, the most important find of Taosi jade was discovered in 2003 while excavating the Middle Taosi period large tomb no. IIM22. Although the tomb had been severely damaged, there were still abundant finds from the undisturbed part of the burial chamber, including a set of 18 jade and stone objects consisting of five *yue* axes, three *qi* axes, one *cong* tube, three sets of *huang* semi-circular ornaments, one set of animal face, and four large green stone cleavers. Quite a few of the *yue* axes also have multicolour lacquered and painted handles. Object types in this tomb match other Taosi tombs found earlier (He 2003). Outside of the Taosi site, *yue* axes, *bi* disks, *cong* tubes, *lianbi* multi-piece disks, *yabi* notched disks, and *huang* semi-circular ornaments were also commonly found in other Taosi Culture cemeteries, including the Xiajin site (Song 2003) and the Qingliangsi site (Shanxi et al. 2002, 2006, 2011). This shows that jade object types were uniform and stable throughout the Taosi Culture.

The Lower Yellow River region used *yue* axes and *dao* bladed objects as ritual objects. The cemetery at the Zhufeng site demonstrates this point very clearly (Cao 2013). Tombs no. M202 and no. M203 are the largest burials at this site. Funerary jade in M202 includes 2 *yue* axes, 1 *dao* bladed object, 1 *guanshi* crown-shaped object, 1 *zan* hairpin, 4 ornamental pendants, 18 ornamental turquoise strands, and more than 980 pieces of thin turquoise. M203 has 3 *yue* axes, 1 *huan* ring, 5 turquoise pendants, and 95 pieces of thin turquoise (Institute of Archaeology Shandong Research Team 1990). The Dantu site in Shandong Province is another site with a high number of jade objects. Thirty-six jade and stone objects have been collected around this site over the course of years, including 18 *yue* axes and 4 *dao* bladed objects. This also reflects the importance and dominant proportion of these two types of jade objects among Longshan jade (Yang 1996). However, what attracted a large amount of scholarly interest were two other types of objects found at Longshan sites in Shandong Province. One is the *gui* tablet with sacred-face motif collected at the Liangchengzhen site, while the other type is the *yazhang* notched tablets collected at the Dafanzhuang site, the Simatai site, the Shangwanjiagou site, and the Luoquanyu site. Scholars used to see these two types of objects as the most important ritual jade in the Longshan Culture in Shandong Province, although alternative ideas have been proposed in recent years. For example, some scholars believe that the *gui* tablet from the Liangchengzhen site might have been produced by the Shijiahe Culture in the Middle Yangtze River region (Zhu 2013).[7] I personally believe that the *yazhang* notched tablet did not originate from Shandong Province (cf. Deng et al. 2014). It most likely originated from central China and was the main ritual jade object of the Xia dynasty, the so-called *xuangui* dark *gui* tablet (Gu and Zhang 2010) (Figure 6.1).[8] Deng has elaborated the symbolic importance of *yazhang* notched tablets in the political life of the Erlitou (or Xia) interaction sphere (Xu 2009; Deng and Wang 2015).

When discussing *yazhang* notched tablets, one must mention jade from northern Shaanxi Province, among which jade excavated from the Shimao site is the most representative. Judging from the large amount of evidence found so far, the most striking types of jade objects in this region are large *dao* bladed objects, *yabi* notched disks, and most importantly *yazhang*

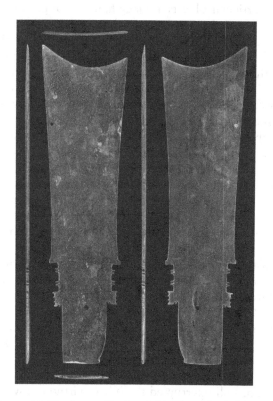

Figure 6.1 *Ya-zhang* tablet (probably *Xuan Gui* in historical records), Erlitou 80VM3:4

notched tablets. According to Deng Shuping's statistics, there are more than a hundred *yazhang* notched tablets from the Shimao site currently in public or private collections in China and abroad, and as such, she believes that Shimao is significant for being the place of origin of the *yazhang* notched tablets (Deng 2014). However, it is important to note that although a large amount of jade had been collected in northern Shaanxi Province (Duan and Zhang 2013), they show very strong external influences. *Cong* tubes from Liangzhu, *yabi* notched disks from Shandong Province, jade animal faces and eagle-shaped *ji* hairpin from the Shijiahe Culture, and *cong* tubes and *bi* disks in the style of the Qijia Culture have all been discovered there. This diversity of exotic jade items was perhaps the result of a lack of a local jade tradition in Shandong. Also taking into account the grandiose walled centre recently found at the Shimao site (Shaanxi et al. 2013), the jade objects at Shimao, including a large amount of *yazhang* notched tablets, were most likely the result of interaction or pillaging instead of local production.

With the onset of the Longshan period and the decline of the Liangzhu Culture, the centre of jade use and production along the Yangtze River shifted upstream to the Shijiahe Culture in the Middle Yangtze River. Current finds of Shijiahe jade[9] mostly come from urn burials at the Xiaojia Wuji site and Tan Jialing site. The most particular objects are jade figurines of gods with fangs and various kinds of animal-shaped objects such as eagles, phoenixes, tigers, and cicadas (Zhang 2008; Figure 6.2). As regards the question of the origins and continuity, although the emergence of Shijiahe jade was quite abrupt, in the sense that the object shapes had not been seen in previous local cultures, such as the Daxi Culture and Qujialing Culture, given that these shapes were also not found in any other contemporary archaeological cultures, we may well consider this a locally developed jade tradition.

By this point, it is possible to witness a significant change in the use of jade during the Longshan period. Jade such as *yue* axes, *dao* bladed objects, and *yazhang* notched tablets that demonstrate worldly power became dominant in the Middle and Lower Yellow River regions, while *cong* tubes and *bi* disks typical to the older tradition were common in the Qijia Culture in Western China. Also, the newly emerged Shijiahe jade was reminiscent of *wu* (shamanistic) jade as most of these jade items are also in mysterious human-animal forms that were definitely beyond everyday activities. In other words, there were now two contrasting jade traditions during the Longshan period, the new *shi* historic jade tradition represented by the Taosi Culture and Shandong Longshan Culture, and the older *wu* (shamanistic) jade tradition represented by the Shijiahe Culture.

In early Chinese society, *wu* (lit. shaman) and *shi* (lit. historian or scribe of records) used to serve the same functions, hence *Discourses of the States: Discourse of Chu II* recorded Guanshefu stating,

> With the fall of Shaohao, Jiuli interrupted the rule of virtuous government. People and gods were mixed up, while name and reality cannot

Figure 6.2 Post-Shijiahe jade – 1. Sacred human mask with fangs, Tanjialing W9:7 (top); 2. Jade eagles, Tanjialing W8:13 (middle); 3. Jade cicada, Tanjialing W8:24 (bottom)

be discerned. Every man held ceremonies; every family proclaimed to be *wu* and *shi*. There was no longer any good faith.

However, *wu* diversified as society developed, dividing into four officials: *zhu*, *zong*, *bu*, and *shi*. *Zhu* was in charge of ceremonies and offerings, *zong* in charge of genealogy, *bu* in charge of divination, and *shi* in charge of records. Even so, early literature often mentions *wu* in tandem with *zhu*, *zong*, *bu*, and *shi*, such as '*wuzhu*', '*wushi*', and '*wubu*', clearly demonstrating their close connection (Li 2000).

The transition from *wu* shamanistic to *shi* historical was admittedly a rationalising process, but in practice, it was also a transition from theocracy to monarchy. The main responsibility of *wu* shamans was to serve the gods and pray for fortune, reflecting the dominance of divine power in society and daily life. However, the responsibility of *shi* historians was 'to reflect upon the relationship between the rules of heaven and the affairs of man' and 'to comprehend the changes of the past and the present' (*Book of Han: Biography of Sima Qian*) according to the grand historian Sima Qian (Ban 1962). A historian should understand providence and nature, but, moreover, he must understand the ways and affairs of man. In other words, the *shi* historian's fundamental role was to assist in governing, therefore the emphasis should be placed on the affairs of man rather than of the gods. Correspondingly, the function of rituals, the main means with which to serve the gods and pray for fortune, became 'to promote filial order and peaceful regeneration, to pacify the country and to settle the people' (*Discourses of the States: Discourse of Chu II*) (Xu 1963). We can say that while the *wu* (shamanistic) jade tradition reflects the ascendancy of divine power, the *shi* historic jade tradition reflects the rise of monarchical power.

6.4 Xia, Shang, and Zhou dynasties: Establishment of the virtuous jade tradition

How the historical Xia dynasty relates to an archaeological culture or period is one of the most debated topics in Chinese archaeology (Liu and Xu 2007). More and more scholars tend to believe that the late Longshan period starts to overlap with the historically recorded Xia dynasty. The most representative argument is the following:

Late Longshan Culture (1900–1750 BC) in Henan Province with the walled city of Wangchenggang as the type site, Xinzhai period remains with Period II at Xinzhai as the type site [cf. Zhao C 2013 for excavation material], and the Erlitou Culture with Erlitou as the type site represent the early, middle, and late stages of the development of the Erlitou Culture, respectively. The walled city of Wangchenggang could have been King Yu's capital called 'Yangcheng'. The Xinzhai site could represented THE Xia Culture during the period when Houyi took over the Xia Dynasty, while Erlitou Culture could represent Xia culture from the Shaokang resurgence to the end of the Xia Dynasty during the reign of King Jie.

(Li 2011)

The Book of Rites: The Record on Example records Confucius saying, 'The Xia Dynasty's way of governing respected the rules of nature and providence. They served spirits and respected gods but kept them at a distance, while they brought the people near and made them loyal' (Dai 2008). This shows that the Xia people already started paying more attention to the affairs of man and distancing themselves from spirits and gods. As discussed in the preceding text, the *shi* historic jade tradition reflecting monarchical power flourished in the Middle and Lower Yellow River regions during the Longshan period. This offers some support to the theory that the late Longshan period overlapped with the Xia dynasty.

If one examines the less controversial Erlitou Culture, clearly *qi* axes and *yazhang* notched tablets were the most important types of jade excavated at the Erlitou site (Yanshi County Museum 1978; Institute of Archaeology 1999b) (Figure 6.3). *Qi* is a subtype of *yue* axe and most likely originated from a production tool resembling axes. However, it is not easy to distinguish whether jade *qi* axes from Erlitou originated from the Taosi Culture or the Shandong Longshan Culture, or even from prehistoric traditions in western Henan Province, because axe type is also found at the Xipo cemetery in Lingbao, for example (Ma et al. 2006). By contrast, the shape of *yazhang* notched tablets is remarkably unprecedented. An object with such particular shape must have been intentionally designed according to the need of some specific rite, therefore it should have originated from one source instead of being developed independently in multiple regions. Based on available evidence, although *yazhang* notched tablets found in western Henan Province are few and later in date, I still believed that *yazhang* notched tablet is the physical form of the most important ritual jade object of the Xia dynasty recorded in literature: the '*xuangui*' (dark *gui* tablet). With the expansion of Xia civilisation, *xuangui* also spread to the so-called four barbarian lands including northern Shaanxi Province, Shandong Province, and the Sichuan Basin (Sun 2013).[10]

As King Tang exiled King Jie of Xia to establish the Shang dynasty, the jade of the Xia dynasty was inherited but also modified. The *yazhang* notched tablet of the Xia dynasty was replaced by the *ge* dagger of the Shang dynasty. While the *yazhang* notched tablet corresponds to the historically recorded *xuangui* insignia (black *gui* tablet) of the Xia dynasty, the *ge* dagger corresponds to the *gui* insignia (*gui* tablet) of the Shang dynasty. The *gui* tablet remained the most important type of ritual object despite dynastic changes, but there was innovation to its specific form as well; this is an example of the change and continuity between Xia and Shang ritual systems. However, although the *yazhang* notched tablet was replace by the

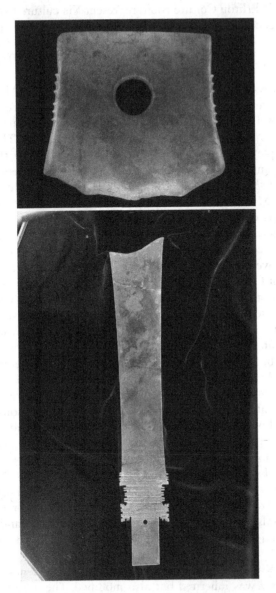

Figure 6.3 Jade from the Erlitou site – 1. *Qi* Axe, Erlitou 81VM6:1 (top); 2. *Ya-zhang*, Erlitou 75VIIIKM7:5 (bottom)

ge dagger in central China, its shape persisted in the Sanxingdui Culture on the Sichuan Basin, to the extent that they substituted the typical Shang-style dagger with notched tablets; this represents an important instance in which lost rites and rituals can still be found outside the main court. More importantly, while the importance of *ge* dagger (or *yu-gui* insignia) was being consolidated, the *yue* axe (or *qi* axe) became obsolete during the Shang dynasty, the former far surpassing the latter both in quantity and in size. Also, *ge* dagger began to reflect hierarchy as well. The more distinguished the owner, the larger the jade dagger. The largest Shang dynasty *ge* dagger discovered so far, 96 cm in length, was excavated at tomb no M2 at the Lijiazui location of the Panlongcheng site, one of the largest burials during the Early Shang period (Hubei 2001a).

In summary, the fact that *gui* tablets (in the form of either *yazhang* notched tablet or *ge* jade dagger) became the dominant type of ritual jade symbolising the development and consolidation of the *shi* historic jade tradition during the Xia and Shang dynasties. At the same time, the ancient *cong* tube, representing the *wu* (shamanistic) jade tradition, gradually vanished. The reason behind the rise of *gui* tablets and the fall of *cong* tubes may be attributed to the dominance of monarchical power over divine power.

The significance of King Wu of Zhou overthrowing the Shang dynasty was not just a change of dynasties, but the fact that it led to the Duke of Zhou's rule and his institutionalisation of rites and music, which set the foundation for the 'rites and music' tradition in Chinese Culture. As rites were institutionalised, ritual objects were also given established shapes. Although the elite had favoured bronze vessels since the beginning of the Xia dynasty, strictly speaking, bronze vessels such as the *ding* tripod, *zu* meat board, *bian* stemmed dish, and *dou* stemmed dish were vessels used to conduct ceremonies, not object offered as gifts to the gods. Jade was still the default material for ritual objects used to pay tribute to the gods and pray for fortune (Sun 2000, 2002, 2003, 2004, 2008).

According to historical records, ritual jade of the Zhou dynasty included the Six Insignia and Six Ritual Objects. *Rites of Zhou: Offices of Spring – Da Zong Bo* records: 'The Six *Rui* Insignia are made with jade to distinguish rank. Kings used *zhen gui* tablet, Dukes used *huan gui* tablet, Marquis used *xin gui* tablet, Counts used *gong gui* tablet, Viscounts used *gu bi* disk, Barons used *pu bi* disk' and

> the Six *Qi* Ritual Objects are made with jade to honour heaven and earth and the four directions. The dark *cang bi* disk was used to honour the heaven, yellow *huang cong* tube was used to honour the earth, green *qing gui* tablet was used to honour the east, red *chi zhang* tablet was used to honour the south, white *bai hu* tiger-shaped ornament was used to honour the west, and black *xuan huang* semicircular ornament was used to honour the north. Sacrificial animals and coins corresponding to the colour of the ritual jade are also dedicated to each of the above.

Rites of Zhou: Offices of Spring – Dian Rui states the responsibilities of the *dian rui* official were 'to manage the collection of the Six Insignia and Six Ritual Objects, to distinguish their type and usage, and to arrange appropriate clothing'. In particular, *Rites of Zhou* mentions 'insignia' and 'ritual objects' separately to distinguish between the two.

Based on the description given in *Rites of Zhou*, the Six Insignia were in fact just two types of objects, *gui* tablets and *bi* disks, while the Six Ritual Objects included six types of objects (*gui* tablet, *zhang* tablet, *bi* disk, *cong* tube, *hu* tiger-shaped ornament, and *huang* semi-circular ornament). Scrutinising the archaeological evidence, one can affirm that *gui* tablets, *zhang* tablets, and *bi* disks were the main types of ritual jade during the Zhou dynasty (Sun 2008). *Gui* tablets had been the main ritual jade object since the Xia dynasty. There are numerous archaeological examples from Zhou dynasty tombs that contain *gui* tablets (Sun 2008). Such jade objects appear in tombs belonging to all kinds of social ranks and in both male and female tombs. *Bi* disks were even more ancient and had been used by many Neolithic cultures. They are the recurrent type of jade in Zhou dynasty tombs. The *zhang* tablet of the Zhou dynasty, recorded both in the written literature and bronze inscriptions, however, is still a mystery as regards to which jade object these inscriptions are referring to. Most scholars believe that the *zhang* tablet refers to what is called the *bingxingqi* handle-shaped object in archaeology (e.g., the Early Western Zhou Yu state cemetery tombs [Lu and Hu 1988]). This *bingxingqi* handle-shaped object originated from the Erlitou period and was used throughout the Shang period. There were two changes to it during the Western Zhou period. The first was the addition of a phoenix-bird motif and dragon motif on the hilt. The second was the addition of turquoise, thin slices of jade, and thin shell inlaid on the front part of the object. Judging from the sophistication of the handle-shaped jade and how widespread it was, certainly its prestige can be compared to that of the *zhang* tablet.

Another important characteristic of the use of jade in the Zhou dynasty was the interchangeable use of *rui* insignia and *qi* ritual objects. *Gui* tablets, *zhang* tablets, and *bi* disks were insignia and formal objects uses by the elite, indispensable in occasions such as when the king appointed vassals, when vassals or nobles presented themselves to the king, and when nobles visited one another. In addition, they were also the most important ritual objects used when paying respect to the gods (Sun 2012). For example, when the Duke of Zhou prayed to the ancestors, he 'laid out *bi* disks and held *gui* tablets' (*Classic of History: Gold Coffer*) (Wang 2011). When King Xuan of Zhou made offerings to heaven, he also had to 'exhaust all *gui* tablets and *bi* disks' (*Book of Songs: Greater Odes – Yun Han*) (Cheng 2006). Therefore, the Han dynasty scholar Zheng Xuan (127–200 AD) explained the difference between *rui* insignia and *qi* ritual objects as: 'That which people hold and demonstrate to others is called *rui*, that which is used to pay respect to gods is called *qi*' (*Zheng's Commentary on 'Rites of Zhou: Offices of Spring – Dian Rui'*) (Zheng 2010). The interchangeability of the use of *rui* insignia and *qi* ritual objects is another instance of ancient people extending their own likings to the gods. In addition, ritual jade found in Zhou period ceremonial sites such as the Houma Mengshi site shows that the types of jade objects far exceeded the types outlined amongst the Six Ritual Objects. For instance, *huan* rings, *huang* semi-circular ornaments, belt hooks, dragon-shape ornaments, human-figure ornaments, and broken jade pieces could all have been presented as gifts to the gods. This shows that the Zhou people were more concerned with jade as a material rather than the shape of objects. Perhaps this is also the key to understanding the statement in *Book of Rites: Single Victim at Border Sacrifices* (Dai 2008):

> That which is most important in ceremonies is to understand the ideas intended in them. While the idea is missed, the number of things and observances in them may be correctly exhibited, as that is the business of the *zhu* officers of prayer and the *shi* historians.

An important development in the Zhou dynasty compared to the Xia and Shang dynasties was the systematisation of the use of jade. There was insignia jade, mainly represented by *gui* tablets, *zhang* tablets, and *bi* disks; ornamental jade, represented by ornaments with multiple *huang* semi-circular ornaments and ornamental strands with trapezoid-shaped plaques; and funerary jade, such as funerary masks. However, what truly represented the sophistication of Zhou jade culture was the association between jade and virtue, from which the tradition of virtuous jade, arguable the epitome of Chinese jade culture, was derived.

Guo Moruo pointed out that *de* (virtue) was an idea particular to the Zhou dynasty. It encompassed both a subjective aspect of self-cultivation and an objective dimension known to posterity as *li* (ritual) (Guo 1982). The Zhou people were renowned for both virtue and filial piety, virtue towards heaven, filial piety towards the ancestors. Therefore 'having both filial piety and virtue' (*Book of Songs: Greater Odes*) (Cheng 2006) was the guiding principle in civilised Zhou society and was a great intellectual progress brought about by the reforms of the Zhou dynasty (Hou et al. 1957). Similarly, Li Zehou believes that *wu* shamanism became divided into two strands during the early Zhou dynasty. 'On the one hand, it developed into the offices of prayer officials, divination officials and recorder/historians, and trickled into folk culture to form a subsidiary tradition'; on the other hand,

> it became rationalised and was institutionalised through the Duke of Zhou's formalisation of rites and music, which systematically preserved and maintained the syncretism between man and heaven and between politics and religion that is fundamental to *wu* shamanism. This became the primary tradition that is the key to Chinese culture. Virtue and rituals represent the completed form of this rationalising process.
>
> (Li 2012)[11]

From the Spring and Autumn period onwards, the humanistic aspect of the Zhou culture of 'rites and music' developed further. The concept of virtue was emphasised like never before. Therefore, the spring and autumn period was in fact already the 'age of virtue' (Chen 2002). The Zhou discussion of virtue was not abstract. Instead, it was tied to a specific group of people known as *junzi* (gentlemen). The concept of *junzi* changed dramatically in the Spring and Autumn period since the moral ideal that it represented was no longer necessarily connected to social status. *Junzi* became a virtuous ideal, hence the explanation in *Virtuous Discussions of the White Tiger Hall: 'Junzi' as a General Term*: 'Who can be called *junzi*? It is a title that denotes virtue' (Ban 1994).

The Book of Rites: Meaning of Exchange between Courts records Confucius saying: '[A]nciently *junzi* compared themselves to jade in terms of virtue' (Dai 2008). What is the connection between *junzi* and jade? The key is that each had their form of virtue. Jade is the most sacred and pure natural substance embodying the virtue of heaven and earth, while *junzi* is the paradigm of human virtue and worldly morality. The two can, therefore, be compared to each other, as in the line of poetry 'I think of my husband, he who is gentle as jade' (*Book of Songs: Odes of Qin – Xiao Rong*) (Cheng 1996). This was how the Chinese virtuous jade tradition originated.

Confucius once exclaimed that 'the simplicity of Yu and Xia and the refinement of Yin and Zhou were both taken to the greatest degree' (*Book of Rites: The Record on Example*) (Dai 2008). With reference to early Chinese jade, the *shi* historic jade culture of the Xia and Shang dynasties and the *de* virtuous jade culture of the Zhou dynasty deserved the remark of 'taken to the greatest degree'. Jade objects and jade culture after the Zhou dynasty were either not as unaffectedly simple as those of the Xia dynasty or not as refined as those of the Shang and Zhou dynasties. Generally speaking, jade culture after the Zhou dynasty declined more than progressed, and the glory of the jade of the Three Dynasties was not seen again. It is fortunate, however, that the Chinese tradition of revering and loving jade was preserved and sustained; that which has always been revered is the virtue of jade. This is an element of Chinese jade culture that is permanent and unaltered.

6.4.1 *Editors' notes*

We briefly discuss why it is important to critically evaluate and cross-reference both historic documents and archaeological materials for a holistic understanding of the jade culture in ancient China. While the burgeoning archaeological campaigns across China have provided a corpus of material evidence to understand the cultural and symbolic role of jade in ancient society, which was not possible previously, the altitudes towards historic documents have been divided. Some of the radical opinions suggest a complete discard of the historical records as most of them, as pointed out by Gu Jiegang in the early twentieth century, were fabricated layers of facts, myths,

and misinterpretations in their transmission. For these scholars, these historical documents introduce unnecessary noise and to get to historical truth,
one should rely solely on archaeological records. This approach has been
criticised for ignoring the historic facts recorded in historical documents.
We argue that the facts about jade in historical documents, if used critically,
could be valuable to understanding of jade cultures in ancient China.

1   We should, of course, treat historical records with caution, and not take
    them at their face value, and certainly not mistake them with historical
    fact. However, we should not ignore their value. Many scholars would
    agree that ethnographic and anthropological data plays an indispens-
    able role in reconstructing ancient history, but very few question the
    credibility of using such data in interpreting the past. Can the present be
    a key to the past? Compared to ethnographic data, we would propose
    that historical documents, despite their often fabricated nature, are near
    to historical facts. Take the book, Rites of Zhou, as an example. It is
    now widely agreed that this book was written during the Warring States
    period (770–221 BC) (Shen and Li 2004), later than the time of the
    Western Zhou period (c.1050–770 BC) rituals dealt with in the book.
    Given this chronological discrepancy, the detailed information of the
    jade rituals in this book reflect the Warring States period's understanding
    of the Western Zhou period. This being said, one should also acknow-
    ledge that without this book, we would not even have known of the
    existence of these rituals. Indeed, without these records, especially their
    descriptions and illustrations, we would probably have still been arguing
    about the naming, function, and many other aspects on the archaeology
    of jade. In this sense, the historical documents should be considered as
    a strength rather than a weakness to the archaeological understanding
    of jade.

2   The important information contained in historical documents should
    not be ignored Not only are there transmitted texts in paper forms
    dating back to as early as the Tang dynasty, but recent excavations have
    also recovered bamboo slips and books made with other rare organic
    materials. These alternative sources of books have also provided
    growing and additional supportive information of early historical
    period classic texts. One recent example is from the excavation of the
    Haihun marquis's mausoleum, in which one of the earliest editions of
    Confucius Analects was preserved and found.

3   Needless to say, these documents should be cross-checked with arch-
    aeological data. The seemingly, well-regulated hierarchy manifested by
    the Western Zhou elite object assemblage, including jade, bronzes, and
    many other items, has frequently been revisited and modified based upon
    archaeological data from the excavation of numerous Bronze Age elite
    tombs. The hierarchical system recorded in the historical documents
    often states one should not enjoy an amount of jade and other items

surpassing their social status. Yet the archaeological evidence often shows that the elites, sometimes the lowest-ranking elites, frequently ignored this regulation by using a number of jade and bronze items that were incomparable with their prescribed social status. These different scenarios illustrated by the archaeological data and historic documents reflects the subtle tension within the elite circle of the Zhou period. The realisation of this subtlety would have not been possible without fruitful dialogue between archaeological data and historical documents. This chapter then adopts this cautious attitude towards historical documents for the archaeology of jade. It is through the combined use of historical documents and archaeological data that we can start to get a handle on the continuous jade culture as a silent characteristic for the material culture of ancient China.

Notes

1 Editors' note: The archaeological search of Xia has been very controversial; see Liu and Xu (2007) for a good overview of the debate and Li (2013) for more information on the history of 'Xia'.
2 Editors' note: The best editions of many of the transmitted texts discussed in this chapter are normally published by either the Zhonghua Book Company or the Shanghai Ancient Works Publishing House. Unless otherwise stated, the transmitted texts discussed in this chapter are cited from books published by these presses. Through their transmission process, scholars in different historical periods would also make their own philological studies of these classics, some of which become the most authoritative edition survived. In such cases, the philological version would be cited.
3 Some scholars have corrected and added to Wang Guowei's explanation of *li* in recent years. For example, Qiu Xigui pointed out that *li* was the name of a kind of drum, possibly a 'large and expensive drum decorated with jade'. Zheng Jiexiang believes that the character *li* corresponded with the character for drum and for double jade (*yu*), referring to 'using jade and the sound of drums and music to conduct ceremonies for heaven, earth, gods and spirits' (Qiu 1980; Zheng 1987).
4 Based on archaeological evidence, some scholars recently insisted that there was a Jade Age in Chinese history, and that jade objects can be seen as a significant symbol of Chinese civilisation (Wu 1990; Mou 1997).
5 Yang Boda thinks instead that '[t]he owners of these excavated prehistoric jade were *wu* shaman only, who used the jade to serve the gods. These jade objects were sacred objects. It seems rather impossible that prehistoric jade was owned by secular and administrative leaders, or used to decorate the body or to represent status in daily life, or to be preserved as a form of wealth' (Yang 2011). However, this argument seems rather too extreme.
6 This sentence is the editors' addition.
7 However, Deng Shuping thinks that it was probably the result of interaction between Longshan Culture and Shijiahe Culture (Deng 2014).
8 Editors' note: All the figures in this chapter are added by the editors.

9 Editors' note: Strictly speaking, the jade items and their archaeological contexts (found in urn burials) belong to the so-called post–Shijiahe Culture.

10 Editors' note: The main reason for professor Sun's adherence to the central-plain origin for this type of important jade might be that we have to understand the symbolic meanings of jade within broader, socio-economic contexts. With the accumulation of amazing archaeological data from the excavations at Erlitou, Xinzhai, and other contemporary late-Longshan-period settlements (Xu 2009; Zhao, C. 2013), most scholars agree that central China is where early states emerged with sophisticated ritual systems and social hierarchy. Thus, it makes sense that *yazhang* notched tablets, which the majority of scholars agree were used in ritual activities (Okamura 2012), and their associated rituals originated from these early complex societies first.

11 Editors' note: Although it is difficult, if not impossible, to substantiate such philosophical and spiritual changes in human minds through the lens of archaeology, the *longue durée* perspective adopted in this chapter offers an important glimpse to this profound transformation in social rituals and underpinning political changes.

Glossary

Anji 安吉
Anxi 安溪
bai hu 白琥 (white tiger-shaped ornament)
bajiaoxing 八角星纹 (octagon)
Banshan 半山
Beinan 卑南
Beiyinyangying 北阴阳营
Beiyinyangying-Lingjiatan 北阴阳营-凌家滩
Beiyinyangying-Sanxingcun 北阴阳营-三星村
ben 锛 (adze)
bian 笾 (stemmed dish)
Biandanshan 扁担山
Biandanshan-Heshangdi 扁担山-和尚地
Bianjiashan 卞家山
bi 璧 (disk)
bingxingqi 柄形器 (handle-shaped object)
Boyishan 钵衣山
bu 卜
Caiyuan 菜园
cang bi 苍璧 (disk)
Caoxieshan 草鞋山
Caoxieshan-Zhaolingshan Site Cluster 草鞋山-赵陵山遗址群
ce 策
Chahai 查海
Changming 长命
Changzhou Site Cluster 常州遗址群
chantou 缠头
Chaoshan 超山
Chen Jie 陈杰
chi zhang 赤璋 (red notched tablet)
Cihu 慈湖
Cong 琮 (tube)
congshizhuo 琮式镯 (*cong*-style bracelet)

congshi zhuxingqi 琮式柱形器 (*cong*-style cylindrical objects)
Dabieshan 大别山
Dadianzi 大甸子
Dafanzhuang 大范庄
Dafen 大坟
daizugui 袋足鬶 (bag-legged kettle)
Damojiaoshan 大莫角山
dang 珰
Dantu 丹土
dao 刀 (bladed object)
dawa 打洼 (create grooves)
Dawenkou 大汶口
Daxi 大溪
Daxiongshan 大雄山
Daxiongshan Site Cluster 大雄山聚落群
Da Yu 大禹
Dazheshan 大遮山
Dazemiao 达泽庙
De 德
Deng Cong 邓聪
Deng Shuping 邓淑蘋
Deqing Site Cluster 德清遗址群
dian rui 典瑞
ding 鼎 (tripod)
Dingshadi 丁沙地
Donglongshan 东龙山
Dongshancun 东山村
Dongshanzhui 东山嘴
Dongtiao River 东苕溪
Dongyangjiacun 东杨家村
dou 豆 (stemmed dish)
duanshi 端饰 (end ornament)
dui 镦 (end ornament)
Duke of Zhou 周公
Dushan 杜山
Erlitou 二里头
Fangjiazhou 方家洲
fangshu 方术
Fanshan 反山
Fengshan 凤山
Fuquanshan 福泉山
Ganggongling 岗公岭
Gansu Province 甘肃省
Gaochengdun 高城墩
Gao Wei 高炜

ge 戈 (dagger)

Gonggong 共工

gong gui 躬圭 (tablet)

gouyunxingqi 勾云形器 (hook-and-cloud ornament)

guan 冠

guanshi 冠饰 (crown-shaped object)

guanzhu 管珠 (tubular bead)

guanzhuangshi 冠状饰 (cockscomb-shaped object)

Guangfulin 广富林

Guanjingtou 官井头

Guanshan 官山

Guanshefu 观射父

Guanzhuang 官庄

gu bi 穀璧 (disk)

Gu Jiegang 顾颉刚

gui 鬶 (kettle)

gui 圭 (tablet)

Gun 鲧

Guo Moruo 郭沫若

Gushan 鼓山

Haihun marquis's mausoleum 海昏侯墓

Haining 海宁

Haiyan 海盐

Haiyan-Pinghu Site Cluster 海盐-平湖遗址群

Hangzhou 杭州

Hangzhou-Jiaxing-Huzhou Plains 杭嘉湖地区

Hemudu 河姆渡

Henan Province 河南省

Hengshan 横山

Heyedi 荷叶地

Hongshan 红山

Houliuhecun 后柳河村

Houma Mengshi 侯马盟誓

Houtoushan 后头山

Houyangcun 后杨村

Houyi 后羿

huan 环 (ring)

huang 璜 (semi-circular ornament)

huang cong 黄琮 (yellow tube)

Huangfentou 皇坟头

huan gui 桓圭 (tablet)

Huating 花厅

Huiguanshan 汇观山

ji 笄 (hairpin)

jia 荚

Jiahu 贾湖
jiandi 减地 (reduce the background)
Jianghuai 江淮
Jiangsu Province 江苏省
Jiangyin 江阴
jianhui 尖喙 (pointy beak-like)
Jiashan 嘉善
Jiaxing 嘉兴
Jie 桀
jieyusha 解玉砂 (abrasive sand)
jigubai 鸡骨白 (chicken-bone white)
Jincun 金村
jishizhong 积石冢 (stone altars)
jue 玦 (slit earring)
jun zi 君子 (gentlemen)
Kaolaoshan 栲栳山
King Wu 武王
King Xuan 宣王
Kuahuqiao 跨湖桥
Lajia 喇家
Lantian shanfang 蓝田山房
Laohuling 老虎岭
li 礼 (ritual)
lianbi 联璧 (multi-piece disk)
Liangchengzhen 两城镇
Liangzhu 良渚
Liangzhu City 良渚古城
Liangzhu Site Cluster 良渚遗址群
Liao River 辽河
lie qin 烈沁
Linping 临平
Linping Site Cluster 临平遗址群
ling 灵 (spirit)
Lin Yun 林沄
Lishan 里山
Lishan-Zhengcun-Gaocun 里山-郑村-高村
Liu Dunyuan 刘敦愿
Liyushan 鲤鱼山
Li Zehou 李泽厚
Longqiuzhuang 龙虬庄
Longshan 龙山
Longtangang 龙潭港
Lucun 卢村
Luoquanyu 罗圈峪
Majiabang 马家浜

Majiayao 马家窑
Majinkou 马金口
mao 瑁 (end ornament)
Maoshan 茅山
Maoyuanling 毛园岭
Maqiao 马桥
matixingqi 马蹄形器 (horse hoof-shaped object)
Meijiali 梅家里
Meiling 梅岭
Meirendi 美人地
Meiyuanli 梅园里
Miaodigou 庙底沟
Miaoqian 庙前
Mifenglong 蜜蜂垄
Mojiaoshan 莫角山
Mopandun 磨盘墩
Mou Yongkang 牟永抗
nanguanghuang 南瓜黄 (pumpkin yellow)
Nanhebang 南河浜
Nanjing City 南京市
Nanshan 南山
Ningbo 宁波
Ningxia Hui Autonomous Region 宁夏回族自治区
Ningzhen 宁镇
Niuheliang 牛河梁
Panlongcheng Lijiazui 盘龙城李家嘴
Pingyao 瓶窑
Pishan 毘山
Puanqiao 普安桥
pu bi 蒲璧 (disk)
Putaofan 葡萄畈
qi 器
qi 戚 (axe)
Qianshanyang 钱山漾
Qiantang River 钱塘江
Qijia 齐家
qing gui 青圭 (green tablet)
Qinghai Province 青海省
Qingliangsi 清凉寺
Qingpu Site Cluster 青浦遗址群
Qiuchengdun 邱承墩
Qiuwu 秋坞
Qiu Xigui 裘锡圭
Qujialing 屈家岭
Rui 瑞

sanchaxingqi 三叉形器 (three-pronged objects)
San Qiao 三蹻
Sanxingdui 三星堆
Shaanxi Province 陕西省
Shandong Province 山东省
Shangkoushan 上口山
Shangwanjiagou 上万家沟
shan liao 山料
Shanxi Province 山西省
Shaokang 少康
Shaoqingshan 少卿山
Shedunmiao 佘墩庙
Sheng Qixin 盛起新
Shenhui 神徽 (insignia)
shenren 神人 (sacred human)
shenren shoumian xiang 神人兽面像 (sacred human and animal-face motif)
shi 史
shicao 蓍草
shifa 蓍法
Shijiahe 石家河
Shijiahe Site Cluster 石家河遗址群
Shijiahe Walled Site 石家河城址
Shimao 石峁
Shin'ichi Nakamura 中村慎一
Shiwu 石坞
Shixia 石峡
Shi Xingeng 施昕更
Shizhaocun 师赵村
Shizishan 狮子山
Shoumian 兽面 (animal-face)
shuangerhu 双耳壶 (two-eared bottle)
Sichuan Basin 四川盆地
Sidun 寺墩
Simao Qian 司马迁
Simatai 司马台
Songze 崧泽
Su Bingqi 苏秉琦
Suzhou 肃州
Tang 汤
Tangshan 塘山
Taosi Walled Site 陶寺城址
Tianluoshan 田螺山
Tianmu 天目山
Tiao River 苕溪
Tinglin 亭林

Tonglin Walled Site 桐林城址
Tonglu 桐庐
Tongxiang 桐乡
Tongxiang-Haining Site Cluster 桐乡-海宁遗址群
tongxingqi 筒形器 (tubular object)
tuoju 砣具 (rotary wheel)
Wangchenggang 王城岗
Wang Guowei 王国维
Wang Mingda 王明达
Wanjiang River 皖江
Wen Guang 闻广
Wenjiashan 文家山
wu 巫
Wuguishan 乌龟山
Wujiabu 吴家埠
Wujiachang 吴家场
Wusong River 吴淞江
Wutongnong 梧桐弄
Wuxian-Kunshan Site Cluster 吴县-昆山遗址群
Wuyang 舞阳
Wu-Yue 吴越
xi 觋
Xiajin 下靳
xiangyabai 象牙白 (ivory white)
xiansou 线搜 (string-cut openwork)
Xiantanmiao 仙坛庙
Xiaodouli 小兜里
Xiaojia Wuji 肖家屋脊
Xiaomeiling 小梅岭
Xiaomojiaoshan 小莫角山
Xieshi 写实 (representational)
Xieyi 写意 (impressionistic)
Xindili 新地里
Xingang 新岗
xin gui 信圭 (tablet)
Xinglongwa-Xinglonggou 兴隆洼-兴隆沟
Xinyi 信宜
Xinzhai 新砦
Xipo 西坡
Xiuyan 岫岩
Xiyangjiacun 西杨家村
xuan gui 玄圭 (tablet)
Xuejiagang 薛家岗
Xunshan 荀山
yabi 牙壁 (notched disk)
Yang Boda 杨伯达

Yangshao 仰韶

Yan Wenming 严文明

Yaojiadun 姚家墩.

Yaojiashan 姚家山

Yaoshan 瑶山

yashiqing 鸭屎青 (duck stool green)

yazhang 牙璋 (notched tablet)

Yinxu 殷墟

yu 珏

yue axe 钺

yuguan 羽冠 (feather crest)

Yuhang District 余杭区

Yujiashan 玉架山

yuntian qi 耘田器 (harvesting tool)

yuzhulong 玉猪龙 (pig-dragon)

zan 簪 (hairpin)

Zanmiao 昝庙

Zhangjiadun 张家墩

Zhanglingshan 张陵山

Zhang Min 张敏

Zhangshan 獐山

zhang 璋 (tablet)

Zhang Zhongpei 张忠培

Zhaolingshan 赵陵山

Zhejiang Province 浙江省

Zheng Jiexiang 郑杰祥

zhen gui 镇圭 (tablet)

Zhili 芝里

Zhishan 雉山

Zhongjiacun 钟家村

Zhongjiashan 仲家山

zhong qi 重器

Zhoucun 周村

Zhoujiafan 周家畈

zhu 祝

Zhuangqiaofen 庄桥坟

Zhuanxu 颛顼

Zhufeng 朱封

zhuixingqi 锥形器 (awl-shaped object)

zhuxingqi 柱形器 (cylindrical object)

zhuoshicong 镯式琮 (bracelet-style *cong* tube)

zhuxinghe 猪形盉 (pig-shaped vessels)

zong 宗

Zongri 宗日

zu 俎 (meat board)

List of historical records

Book of Han: Biography of Sima Qian (Han Shu: Simao Qian Zhuan 汉书·司马迁传)

Book of Rites: Meaning of Exchange between Courts (Li ji: Pin Yi 礼记·聘义)

Book of Rites: Single Victim at Border Sacrifices (Liji: Jiao Te Sheng 礼记·郊特牲)

Book of Rites: The Record on Example (Li Ji: Biao Ji 礼记·表记)

Book of Rites by the Older Dai: Virtue of the Five Kings (Da Dai Li Ji: Wu Di De 大戴礼记·五帝德)

Book of Songs: Greater Odes – Juan A (Shi Jing: Da Ya – Juan A 诗经·大雅·卷阿)

Book of Songs: Greater Odes – Yun Han (Shi Jing: Da Ya – Yun Han. 诗经·大雅·云汉)

Book of Songs: Odes of Qin – Xiao Rong (Shi Jing: Qin Feng – Xiao Rong 诗经·秦风·小戎)

Classic of History: Gold Coffer (Shang Shu: Jin Teng 尚书·金滕)

Confucius Analects: Wei Zheng (Lun Yu: Wei Zheng 论语·为政)

Discourses of the States: Discourse of Chu II (Guoyu: Chuyu Xia 国语·楚语下)

*Exemplary Sayings: Chong Li (Fa Yan: Chong L i*法言·重黎)

Explaining Graphs and Analysing Characters (Shuowen Jiezi 说文解字)

Explaining Graphs and Analysing Characters: Jade Section (Shuowen Jiezi: Yu Bu 说文解字·玉部)

Lv's Annals: Almanac of the Third Month of Autumn (Lv Shi Chun Qiu: Ji Qiu Ji 吕氏春秋·季秋纪)

Records of the Grand Historian: Book of Rites (Shi ji: Li Shu 史记·礼书)

Rites of Zhou (Zhou Li 周礼)

Rites of Zhou: Offices of Spring – Da Zong Bo (Zhou li: Chun Guan – Da Zong Bo 周礼·春官·大宗伯)

Rites of Zhou: Offices of Spring – Dian Rui (Zhou Li: Chun Gguan – Dian Rui 周礼·春官·典瑞)

Virtuous Discussions of the White Tiger Hall: 'Jun zi' as a General Term (Bai Hu Tong: Jun Zi Wei Tong Cheng 白虎通·君子为通称)

Zheng's Commentary on 'Rites of Zhou: Offices of Spring – Dian Rui (Zhou Li: Chun Guan – Dian Rui 郑注周礼·春官·典瑞)

Bibliography

Adams, R. M. (1981). *Heartland of Cities: Surveys of Ancient Settlement and Land Use on the Central Floodplain at the Euphrates*. Chicago: University of Chicago Press.

Algaze, G. (2001). The Prehistory of Imperialism: The Case of Uruk Period Mesopotamia, in M. S. Rothman (ed.), *Uruk Mesopotamia and Its Neighbors: Cross-Cultural Interactions in the Era of State Formation*. Santa Fe, NM: School of American Research Press, pp. 27–83.

Algaze, G. (2005). *The Uruk World System: The Dynamics of Expansion of Early Mesopotamian Civilization*, 2nd rev. ed. Chicago: University of Chicago Press.

Algaze, G. (2009). *Ancient Mesopotamia at the Dawn of Civilization: The Evolution of an Urban Landscape*. Chicago: University of Chicago Press.

Anhui Provincial Institute of Cultural Relics and Archaeology 安徽省文物考古研究所. (2004). *Qianshan Xuejiagang 潜山薛家岗*. Beijing: Cultural Relics Press.

———. (2006). *Lingjiatan-Tianye Kaogu Fajue Baogao Zhiyi 凌家滩—田野考古发掘报告之一 (Lingjiatang: The Archaeological Excavation Report, Vol. 1)*. Beijing: Cultural Relics Press.

———. (2008). Anhui Hanshanxian lingjiatan yizhi diwuci fajue de xinfaxian 安徽含山县凌家滩遗址第五次发掘的新发现 (The New Discoveries in the Fifth Excavation of the Lingjiatan Site, Hanshan County, Anhui). *Kaogu考古 (Archaeology)* (3): 7–17 and plates 1–7.

Anhui Provincial Institute of Cultural Relics and Archaeology 安徽省文物考古研究所 (ed.). (2006). *Lingjiatan Wenhua Yanjiu 凌家滩文化研究 (Studies on the Lingjiatan Culture)*. Beijing: Cultural Relics Press.

Armstrong, K. (2010). *Zhouxin Shidai 轴心时代 (The Great Transformation: The Beginning of Our Religious Traditions)*, Y. Y. Sun and Y. B. Bai (trans.). Haikou: Hainan Press.

Baines, J. (1995). Origins of Egyptian Kingship, in D. B. O'Connor, D. P. Silverman (eds.), *Ancient Egyptian Kingship (Vol. 9)*. Leiden, The Netherlands: Brill, pp. 95–156.

———. (2014). Civilization and Empires: A Perspective on Erligang from Early Egypt, in K. Steinke, D .C. Ching (eds.), *Art and Archaeology of the Erligang Civilization*. Princeton, NJ: Princeton University Press, pp. 99–119.

Ban, G. 班固, and Yan, S. G. 颜师古 (1962). *Han Shu 汉书 (Book of Hou Han)*. Beijing: Zhonghua Book.

Ban, G., et al. 班固等 (1994). *Baihutong Shuzheng 白虎通疏证 (Annotations of Bai Hu Tong)*. Beijing: Zhonghua Book.

Bunbury, J. (2018). Habitat Hysteresis in Ancient Egypt, in Y. Zhuang, M. Altaweel (eds.), *Water Societies and Technologies from the Past and Present*. London: UCL Press, pp. 40–61.

Cao, F. F. 曹芳芳 (2013). Shandong Longshan Wenhua Yongyu Zhidu De Kaoguxue Kaocha 山东龙山文化用玉制度的考古学考察 (Archaeological Research of the Jade Use System of Shandong Longshan Culture). *Yuqi Kaogu Tongxun* 玉器考古通讯 *(Correspondence of Archaeology of Jade)* (2): 60–86.

———.(2014).Longshan Shidai Yongyu Zhidu yu Chuantong de Shanbian: Yi Huanghe Liuyu wei Zhongxin 龙山时代用玉制度与传统的嬗变: 以黄河流域为中心 (Jade Use Systems and Changes during the Longshan Period, Based on the Yellow River Region). Master's thesis, School of Archaeology and Museology, Peking University.

Changzhou Museum 常州博物馆. (2012). *Changzhou Xingang: Xinshiqi Shidai Wenhua Yizhi Fajue Baogao* 常州新岗: 新石器时代文化遗址发掘报告 *(Changzhou Xingang: Excavation Report of the Neolithic Site)*. Beijing: Cultural Relics Press.

Chen, J. 陈杰 (2014). Liangzhu shehui de quanli jiegou 良渚社会的权力结构 (Power Structure in Liangzhu Society), in Shanghai Museum 上海博物馆 (ed.). '*Chengshi yu Wenming*' *Guoji Xueshu Yantaohui Lunwengao* '城市与文明' 国际学术研讨会论文稿 *(Papers from the International Conference 'Cities and Civilisations')* Shanghai: Shanghai Ancient Books, pp. 276–290.

Chen, L. 陈来 (1996). *Gudai Zongjiao yu Lunli: Rujia Sixiang de Genyuan* 古代宗教与伦理— 儒家思想的根源 *(Ancient Religion and Ethics: The Origin of Confucianism)*. Beijing: Science Press.

———. (2002). *Gudai Sixiang Wenhua de Shijie: Chunqiu Shidai de Zongjiao Lunli yu Shehui Sichao* 古代思想文化的世界: 春秋时代的宗教、伦理与社会思潮 *(The World of Ancient Thoughts and Culture: Religion, Ethics and Ethos of the Spring and Autumn Period)*. Hong Kong: Sanlian Publishing.

Chen, X. C. 陈星灿 (2013). Miaodigou shidai: Zaoqi zhongguo wenming de diyilv shuguang 庙底沟时代: 早期中国文明的第一缕曙光 (Miaodigou Period: The Dawn of Early Chinese Civilisation). *Zhongguo Wenwubao* 中国文物报 *(China Cultural Relics News)*, 21 June, Section 5.

Cheng, J. Y. 程俊英 (2006). *Shi Jing* 诗经 *(Book of Songs)*. Shanghai: Shanghai Ancient Books.

Dai , D. 戴德 (2005). Wu di de 五帝德 (The Virtues of the Five Emperors), in H. X. Huang (ed.), *Dadai Liji Huijiao Jizhu* 大戴礼记汇校集注 *(The Collection of Comments and Punctuation on the Book of Rites by the Older Dai)*. Xi'an: San Qin Publishing, pp. 722–776.

Dai, S. 戴圣 (2008). Liji Pingyi 礼记聘义 (The Meaning of the Interchange of Missions between Different Courts), in Y. R. Lv (ed.), *Liji Zhengyi* 礼记正义 *(The Notes and Commentaries of the Book of Rites)*. Shanghai: Shanghai Ancient Books.

Deng, C. 邓聪 (2005). Yiroukegang: Shasheng jieyu kao 以柔克刚: 砂绳截玉考 (Overcoming the Hard with the Soft: Research on Using String and Sand to Cut Jade). *Gugong Wenwu Yuekan* 故宫文物月刊 *(Cultural Relics of Palace Museum)* 265(1): 1–31.

Deng, C. 邓聪, Wang, F. 王方 (2015). Erlitou yazhang (VM3:4) zai nanzhongguo de boji: Zhongguo zaoqi guojia zhengzhi zhidu qiyuan he kuosan 二里头牙璋 （VM3：4）在南中国的波及—中国早期国家政治制度起源和扩散 (Spread of Erlitou Yazhang Jade Tablet [VM3:4] in South China: Origin and Dispersal of Early State Politic Order in China). *Zhongguo Guojia Bowuguan Guankan* 中国国家博物馆馆刊 *(Journal of National Museum of China)* (5): 6–22.

Deng, C., et al. 邓聪等 (2014). Dongya zuizao de yazhang 东亚最早的牙璋: 山东龙山式牙璋初论 (The Earliest Yazhang Jade Tablets in East Asia: An Introduction to the Longshan-Style Yazhang in Shandong), in C. Deng, F. S. Luan, and Q. Wang (eds.), *Yu Rui Dongfang: Dawenkou-Longshan-Liangzhu Yuqi Wenhua Zhan* 玉润东方: 大汶口-龙山-良渚玉器文化展 *(Exhibition of the Dawenkou – Longshan-Liangzhu Jade and Culture)*. Beijing: Cultural Relics Press, pp. 51–62.

Deng, S. P. 邓淑蘋 (1986). Gudai yuqi shang qiyi wenshi de yanjiu 古代玉器上奇异纹饰的研究 (Study on Strange Motifs on Ancient Jade). *Gugong Xueshu Jikan* 故宫学术季刊 *(The National Palace Museum Research Quaterly)* 4: 2–3.

———. (1995). Zhongguo gudai yuqi wenhua sanyuan lun 中国古代玉器文化三源论 (Theory of the Three Origins of Ancient Chinese Jade Cultures). *Zhonghua Wenwu Xuehui Niankan* 中华文化学会年刊 *(Annual Bulletin of the Association of Chinese Culture)*: 197–212.

———. (2007). 'Huaxi xitong yuqi' guandian de xingcheng yu yanjiu zhanwang '华西系统玉器'观点形成与研究展望 (Formation of the 'Western Chinese Jade System' and Research Prospects). *Gugong Xueshu Jikan* 故宫学术季刊 *(The National Palace Museum Research Quarterly)* 25: 1–54.

———. (2014). Wangbang yubo: Xia wangchao de wenhua diyun 万邦玉帛: 夏王朝的文化底蕴 (Jade and Silk of All Kingdoms: Cultural Core of the Xia Dynasty), in H. Xu 许宏 (ed.), *Xiashang Duyi yu Wenhua* 夏商都邑与文化 *(Cities and Culture during the Xia and Shang Dynasties)*, Vol. 2. Beijing: China Social Science Press, pp. 145–246.

Ding, P. 丁品, Zheng, Y. F. 郑云飞 (2010). Zhejiang Yuhang Linping Maoshan Yizhi 浙江余杭临平茅山遗址 (Maoshan Site at Linping, Yuhang, Zhejiang). *Zhongguo Wenwubao* 中国文物报 *(China Cultural Relics News)*. 12 March.

Duan, S. Y. 段双印, Zhang, H. 张华 (2013). Yan'an Chutu Shiqian Yuqi Zonghekaocha yu Yanjiu 延安出土史前玉器综合考察与研究 (Comprehensive Survey and Research on Prehistoric Jade Excavated at Yan'an). *Yuqi Kaogu Tongxun* 玉器考古通讯 *(Correspondence of Archaeology of Jade)* (2): 15–30.

Fang, X. M. 方向明 (2006). Lingjiatan yizhi chutu yuqi xing he wenshi de xiangguan wenti taolun 凌家滩遗址出土玉器形和纹饰的相关问题讨论 (Discussions on the Shapes and Ornamentations of the Lingjiatan Jade Objects), in Anhui Provincial Institute of Cultural Relics and Archaeology (ed.), *Lingjiatan Wenhua Yanjiu* 凌家滩文化研究 *(Studies on the Lingjiatan Culture)*. Beijing: Cultural Relics Press, pp. 190–201.

———. (2013a). Shiqian Zhuoyu de Qiege Gongyi 史前琢玉的切割工艺 (Cutting Techniques in Prehistoric Jade Working). *Nanfang Wenwu* 南方文物 *(Cultural Relics of South China)* (4): 57–61 and 70.

———. (2013b). Liangzhu wenhua de shenren shoumian xiang: Yuqi shidai guannian xingtai he meishu xingshi de ge'an yanjiu 良渚文化的神人兽面像: 玉器时代观念形态和美术形式的个案研究 (The Sacred Man and Animal Mask Motif in the Liangzhu Culture: A Case Study of Ideology and Art Form during the 'Jade Age'), in G. Z. Chen 陈光祖, Z. H. Zang 臧振华 (eds.), *Dongya Kaogu de Xinfaxia: Disijie Zhongyang Yanjiuyuan Guoji Hanxue Huiyi Lunwenji*

东亚考古的新发现：第四届中央研究院国际汉学会议论文集 *(New Discoveries in East Asian Archaeology: Proceedings of the Fourth Academia Sinica International Sinology Conference)*. Taipei: Academia Sinica, pp. 153–198.

———. (2015). Juluo bianqian he tongyi xinyang de xingcheng: Cong songze dao liangzhu 聚落变迁和统一信仰的形成：从崧泽到良渚 (Settlements Changes and the Formation of a Unified Belief: From Songze to Liangzhu). *Dongnan Wenhua* 东南文化 *(Southeast Culture)* (1): 102–112.

Flad, R. (2018). Urbanism as Technology in early China. *Archaeological Research in Asia* 14: 121–134.

Frachetti, M. D., et al. (2012). Multiregional Emergence of Mobile Pastoralism and Nonuniform Institutional Complexity across Eurasia. *Current Anthropology* 53: 2–38.

Gan, F., et al. 干福熹等 (2011). Zhejiang yuhang liangzhu yizhiqun chutu yuqi de wusun fenxi yanjiu 浙江余杭良渚遗址群出土玉器的无损分析研究 (Non-Destructive Analysis of Jade Excavated from the Liangzhu Site Cluster in Yuhang, Zhejiang). *Zhongguo Kexue: Jishu Kexue* 中国科学：技术科学 *(Scientia Sinica [Technologica])* 41(1): 1–15.

Gao, W. 高炜 (1998). Taosi wenhua yuqi ji xiangguan wenti 陶寺文化玉器及相关问题 (Taosi Jade and Related Issues), in C. Deng, 邓聪 (ed.). *Dongya Yuqi* 东亚玉器 *(East Asian Jade)*, Vol. 1. Hong Kong: Centre for Chinese Archaeology and Art of the Chinese University of Hong Kong, pp. 192–200.

Garbrecht, G. (1985). Sadd-el-Kafara: The World's Oldest Large Dam. *Water Power and Dam Construction* 27: 71–76.

Gibson, J. L. (2001). *The Ancient Mounds of Poverty Point: Place of Rings*. Gainesville: University Press of Florida.

Giddens, A. (1986). *Constitution of Society: Outline of the Theory of Structuration*. Cambridge: Polity Press.

Giddy, L. L., Jeffreys, D. G. (1991). Memphis, 1990. *The Journal of Egyptian Archaeology* 77: 1–6.

Goodman, L. E. (1981). *Monotheism: A Philosophic Inquiry into the Foundations of Theology and Ethics*. Lanham, MD: Rowman and Littlefield.

Gu, D. H. 顾冬红 (2009). Jiangyin Gaochengdun yizhi chutu yuqi de jiance he fenxi baogao 江阴高城墩遗址出土玉器的检测和分析报告 (Analytical Report of Jade Objects Excavated from Gaochengdun Site in Jiangyin), in Nanjing Museum 南京博物院 and Jiangyin Museum 江阴博物馆 (eds.), *Gaochengdun* 高城墩. Beijing: Cultural Relics Press.

Gu, F. 古方 (ed.). (2005). *Zhongguo Chutu Yuqi Quanji* 中国出土玉器全集 *(Complete Collection of Jade Excavated in China)*, Vol. 15. Beijing: Science Press.

Gu, F., Li, H. J. 李红娟 (2009). *Guyu de Yuliao* 古玉的玉料 *(Raw Materials for Ancient Jade)*. Beijing: Cultural Relics Press.

Gu, W. F. 顾万发, Zhang, S. L. 张松林 (2010). Lun huadizui yizhi suochu mo yuzhang 论花地嘴遗址所出墨玉璋 (Discussion on the Dark Jade *Zhang* Tablet Found at Huadizui), in Zhengzhou Institute of Cultural Relics and Archaeology 郑州市文物考古研究院 (ed.) *Zhengzhou Wenwu Kaogu yu Yanjiu (II)* 郑州文物考古与研究 (二) *(Archaeology and Research of Zhengzhou II)*. Beijing: Science Press, pp. 802–809.

Guo, D. S. 郭大顺 (1997). Hongshan wenhua de 'weiyuweizang' yu liaohe wenming qiyuan tezheng zairenshi 红山文化的 '惟玉为葬' 与辽河文明起源特征再认识 (A Reappraisal of the Practice of 'Burial with Jade Only' in the Hongshan Culture and Characteristics of the Origin of the Liaohe Civilisation). *Wenwu* 文物 *(Cultural Relics)* (8): 20–26.

———. (1998). Hongshan wenhua yuqi tezheng jiqi shehui wenhua yiyi zai renshi 红山文化玉器特征及其社会文化意义再认识 (Reappraisal of the Characteristics of the Hongshan Culture Jade and Social Implications), in C. Teng, 邓聪 (ed). *Dongya Yuqi* 东亚玉器 *(East Asian Jade)*, Vol. 1. Hong Kong: Centre for Chinese Archaeology and Art of the Chinese University of Hong Kong, pp. 140–147.

Guo, M. R. 郭沫若 (1982). Qingtong Shidai: Xianqin Tiandaoguan de Jinzhan 青铜时代·先秦天道观的进展 (The Bronze Age: The Cosmological Development during the Pre-Qin Era), in *Guo Moruo Quanji: Lishibian* 郭沫若全集·历史编 *(Complete Words of Guo Moruo: History)*, Vol. 1. Beijing: People's Press, pp. 317–376.

Han, J. Y. 韩建业 (2012). Miaodigou shidai yu 'Zaoqi Zhongguo' 庙底沟时代与' 早期中国' (The Miaodigou Period and Early China). *Kaogu* 考古 *(Archaeology)* (3): 59–69.

———. (2015). *Zaoqi Zhongguo: Zhongguo Wenhuaquan de Xingcheng he Fazhan* 早期中国：中国文化圈的形成和发展 *(Early China: The Formation and Development of Chinese Cultural Circle)*. Shanghai: Shanghai Ancient Books.

He, N., et al. 何驽等 (2003). Taosi chengzhi faxian taosi wenhua zhongqi muzang 陶寺城址发现陶寺文化中期墓葬 (Middle Taosi Period Burials Found in the Taosi City). *Kaogu* 考古 *(Archaeology)* (9): 3–6.

———. (2009). Shanxi xiangfen taosi chengzhi zhongqi wangji damu IM22 chutu qigan 'guichi' gongneng chutan 山西襄汾陶寺城址中期王级大墓IM22出土漆杆'圭尺'功能初探 (On the Function of the Guichi Lacquer Stick from the Middle Period Royal Tomb IM22 at the Taosi Walled Town. *Kaogu* 考古 *(Archaeology)* 28: 261–276.

———. (2013). The Longshan Period Site of Taosi in Southern Shanxi Province, in A. Underhill (ed.), *A Companion to Chinese Archaeology*. Hoboken, NJ: Wiley, pp. 255–277.

Helms, S. W. (1981). *Jawa, Lost City of the Black Desert*. Ithaca, NY: Cornell University Press.

Henan Provincial Institute of Cultural Relics and Archaeology. (1999). *Wuyang Jiahu* 舞阳贾湖 *(Wuyang Jiahu Site)*. Beijing: Science Press.

Hou, W. L., et al. 侯外庐等 (1957). *Zhongguo Sixiang Tongshi* 中国思想通史 *(Comprehensive Intellectual History of China)*, Vol.1. Beijing: People's Press.

Huang, J. Q. 黄建秋 (2011). Huangting mudi de yanjiu 花厅墓地研究 (Research on the Huating Cemetery). *Huaxia Kaogu* 华夏考古 *(Huaxia Archaeology)* (3): 64–80.

Huang, Y. P. 黄宜佩 (1987). Luelun woguo xinshiqi shidai yuqi 略论我国新石器时代玉器 (Brief Discussion of Neolithic Jade in China). *Shanghai Bowuguan Jikan* 上海博物馆集刊 *(Bulletin of the Shanghai Museum)* (4): 150–170.

Hubei Provincial Institute of Cultural Relics and Archaeology 湖北省文物考古研究所. (2001a). *Panlongcheng: 1963 nian – 1994 nian Kaogu Fajue Baogao* 盘龙城：1963年~1994年考古发掘报告 *(Panlongcheng: Archaeological Report of Excavations in 1963–1994)*. Beijing: Cultural Relics Press.

———. (2001b). *Wuxue Gushan: Xinshiqi Shidai Mudi Fajue Baogao* 武穴鼓山：新石器时代墓地发掘报告 *(Gushan Cemetery at Wuxue: An Excavation Report on The Neolithic Cemetery)*. Beijing: Science Press.

Hung, L. Y. (2011). Pottery Production, Mortuary Practice, and Social Complexity in the Majiayao Culture, NW China (ca. 5300-4000 BP). PhD Dissertation, Washington University in St. Louis.

Institute of Archaeology, Chinese Academy of Social Sciences 中国社会科学院考古研究所. (1999a). *Shizhaocun yu Xishanping* 师赵村与西山坪 *(Shizhaocun and Xishanping Sites)*. Beijing: Encyclopaedia of China Publishing.

———. (1999b). *Yanshi Erlitou* 偃师二里头. Beijing: Encyclopedia of China Publishing.

Institute of Archaeology, Chinese Academy of Social Sciences 中国社会科学院考古研究所. (1996). *Dadianzi: Xiajiadian Xiaceng Wenhua Yizhi yu Mudi Fajue Baogao* 大甸子: 夏家店下层文化遗址与墓地发掘报告 *(Dadianzi: Excavation Report of Lower Xiajiadian Site and Cemetery)*. Beijing: Science Press.

Institute of Archaeology Gansu-Qinghai Research Team, Chinese Academy of Social Sciences 中国社会科学院考古研究所甘青工作队. (2002). Qinghai minhe lajia yizhi 2000 nian fajue jianbao 青海民和喇家遗址2000年发掘简报 (Brief Report of the Excavation of the Lajia Site in Minhe, Qinghai, 2000). *Kaogu 考古 (Archaeology)* (12): 12–28.

———. (2004). Qinghai minhe lajia yizhi faxian qijia wenhua jitan he ganlanshi jianzhu 青海民和喇家遗址发现齐家文化祭坛和干栏式建筑 (Qijia Culture Altars and Stilt Houses Discovered at the Lajia Site, Minhe, Qinghai). *Kaogu 考古 (Archaeology)* (6): 3–6.

Institute of Archaeology Shandong Research Team, Chinese Academy of Social Sciences 中国社会科学院考古研究所山东工作队. (1990). Shandong linqu zhufeng longshan wenhua muzang 山东临朐朱封龙山文化墓葬 (Longshan Burials at Zhufeng, Linqu, Shandong). *Kaogu 考古 (Archaeology)* (7): 587–594.

Jarzombek, M. M. (2014). *Architecture of First Societies: A Global Perspective*. Hoboken, NJ: Wiley.

Jeffreys, D., Tavares, A. (1994). The Historic Landscape of Early Dynastic Memphis. Mitteilungen des Deutschen Archäologischen Instituts. *Abteilung Kairo* 50: 143–173.

Jiang, S. H. 蒋素华 (2005). Wudi zaoqi yuqi yu meiling yukuang de guanxi 吴地早期玉器与梅岭玉矿的关系 (Relationship between Early Jade from the Taihu Region and the Jade Mine at Meiling). *Wenwu Baohu yu Kaogu Kexue* 文物保护与考古科学 *(Conservation of Cultural Relics and Archaeological Science)* 17: 50–54.

Jiangsu Provincial Institute of Archaeology of Nanjing Museum 南京博物院江苏省考古研究所 and Wuxi Municipal Committee of the Management of Cultural Relics of the Xishan District 无锡市锡山区文物管理委员会. (2010). *Qiuchengdun* 邱承墩 *(The Qiuchengdun Site)*. Beijing: Science Press.

Jingzhou Museum 荆州博物馆, Hubei Provincial Institute of Cultural Relics and Archaeology 湖北省文物考古研究所, and Shijiahe Archaeological Team of School of Archaeology and Museology, Peking University 北京大学考古学系石家河考古队. (1999). *Xiaojia Wuji* 肖家屋脊 *(The Qiuchengdun Site)*. Beijing: Cultural Relics Press.

Kenoyer, J. M. (2000). *Ancient Cities of the Indus Valley Civilisation*. C. X. Zhang (trans.). Hangzhou: Zhejiang People's Press.

Kenoyer, J. M., Heuston, K. B. (2005). *The Ancient South Asian World*. Oxford: Oxford University Press.

Kidder, T. R. (2002). Mapping Poverty Point. *American Antiquity* 67: 89–101.

Li, B. Q. 李伯谦 (2011). Preface, in *Wenming Tanyuan yu Sandai Kaogu Lunji* 文明探源与三代考古论集 (*Collected Essays on the Origin of Civilisation and Archaeology of the Three Dynasties*). Beijing: Cultural Relics Press, pp. 1–4.

———. (2016). Cong zhongguo wenminghua lichen yanjiu kan guojia qiyuan de ruogan lilun wenti从中国文明化历程研究看国家起源的若干理论问题 (Some Theoretical Issues in Understanding the Origin of States from the Study of the Emergence of Chinese Civilization). *Zhongyuan Wenhua Yanjiu中原文化研究 (Studies of Central Plain Culture)* (1): 5–9.

Li, F. (2013). Early China: A Social and Cultural History. Cambridge: Cambridge University Press.

Li, L. 李零 (2000). Xianqinlianghan wenzi ziliaozhong de wu: Shang, Xia 先秦两汉文字资料中的 '巫': 上、下 (*Wu* Shaman in the Literary Records of the Pre-Qin and Han Periods, Parts I and II), in *Zhongguo Fangshu Xukao* 中国方术续考 (*Further Research on Chinese Necromancy*). Shanghai: Dongfang Chubanshe (Orient Publishing).

Li, Z. 李泽厚 (2012). *Shuo Wu Shi Chuantong* 说巫史传统 (*On the Tradition of Wu and Shi*). Shanghai: Shanghai Translation Publishing.

Liaoning Provincial Institute of Cultural Relics and Archaeology 辽宁省文物考古研究所. (2012). *Niuheliang: Hongshan Wenhua Yizhi Fajue Baogao 1983–2003 Niandu牛河梁: 红山文化遗址发掘报告, 年度 (Niuheliang: Excavation Report of a Hongshan Culture Site in 1983–2003)*. Beijing: Cultural Relics Press.

Lin, Y. 林沄. (1965). Shuo Wang 说王 (Explaining *Wang*, the Character for King). *Kaogu 考古 (Archaeology)* (6): 311–312.

Liu, B. 刘斌 (2001). Songze wenhua yu liangzhu wenhua yuqi de bijiao yanjiu 崧泽文化与良渚文化玉器的比较研究 (Comparative Research on Songze and Liangzhu Jade), in X. Qian et al. 钱宪和等 (eds.), *Haixia Liangan Guyuxue Huiyi Lunwenji 海峡两岸古玉学会议论文集 (Collected Essays of the Conference on the Study of Ancient Jade in Mainland China and Taiwan)*. Taipei: Department of Geological Sciences, National Taiwan University, pp. 213–220.

———. (2006). Liangzhu wenhua de jitan yu guanxiang cenian 良渚文化的祭坛与观象测年 (Altar and Astronomical Dating in the Liangzhu Culture), in Zhejiang Provincial Institute of Cultural Relics and Archaeology (ed.), *Zhejiangsheng Wenwu Kaogu Yanjiusuo Xuekan Dibaji: Jinian Liangzhu Yizhi Faxian 70 Zhounian Xueshu Yantaohui Wenji* 浙江省文物考古研究所学刊第八辑—纪念良渚遗址发现70周年学术研讨会文集 *(Eighth Academic Periodical of the Zhejiang Provincial Institute of Cultural Relics and Archaeology: Collected Essays from the Conference Commemorating the 70th Anniversary of the Discovery of the Liangzhu Site)*. Beijing: Science Press.

———. (2007). *Shenwu de Shijie: Liangzhu Wenhua Zonglun* 神巫的世界: 良渚文化综论 (*The World of Gods and Shamans: A Comprehensive Review of the Liangzhu Culture*). Hangzhou: Zhejiang Photography Press.

———. (2013). *Shenwu de Shijie* 神巫的世界 (*The World of Gods and Priests*). Hangzhou: Hangzhou Press.

Liu, B., et al. (2017). The Earliest Hydraulic Enterprise in China, 5,100 Years Ago. *Proceedings of the National Academy of Sciences* 114 (52): 13637–13642.

Liu, D. Y. 刘敦愿 (1972). Ji Liangchengzhen yizhi faxian de liangjian shiqi 记两城镇遗址发现的两件石器 (On Two Stone Objects Discovered at Liangchengzhen). *Kaogu 考古 (Archaeology)* (4): 56–57.

———. (1988). Youguan Rizhao liangchengzhen yukeng yuqi de ziliao 有关日照两城镇玉坑玉器的资料 (Information on the Jade Objects Found in the Jade Pit at Liangchengzhen, Rizhao). *Kaogu 考古 (Archaeology)* (2): 121–123.

———. (2012). Dawenkou wenhua taoqi yu zhubian yishu 大汶口文化陶器与竹编艺术 (Pottery from the Dawenkou Culture and the Art of Bamboo Weaving), in *Liu Dunyuan Wenji 刘敦愿文集 (Collected Works of Liu Dunyuan)*. Beijing: Science Press, pp. 108–112.

Liu, L. (2003). 'The Products of Minds as Well as of Hands': Production of Prestige Goods in the Neolithic and Early State Periods of China. *Asian Perspectives* 42: 1–40.

Liu, L., Chen, X. C. (2012). *The Archaeology of China: From the Late Paleolithic to the Early Bronze Age*. Cambridge: Cambridge University Press.

Liu, L., Xu, H. (2007). Rethinking Erlitou: Legend, History and Chinese Archaeology. *Antiquity* 81: 886–901.

Lou, H., et al. 楼航等 (2012). Zhejiang yuhang yujiashan yizhi faxianle you liuge xianglin de huanhao zucheng de liangzhu wenhua wanzheng juluo 浙江余杭玉架山遗址发现了由六个相邻的环壕组成的良渚文化完整聚落 (The Yujiashan Site at Yuhang, Zhejiang: Discovery of a Complete Liangzhu Settlement Composed of Six Moated Residential Areas). *Zhongguo Wenwubao 中国文物报 (China Cultural Relics News)*, 24 February, Section 4.

Lu, L. C. 卢连成, Hu, Z. S. 胡智生 (1988). *Baoji Yuguo Mudi 宝鸡渔国墓地 (The Royal Cemetery of the Yu State in Baoji)*. Beijing: Cultural Relics Press.

Lu, X. K. 鲁晓珂, Li, W. D. 李伟东, Liu, B. 刘斌, Li, X. W. 李新伟 (2013). Liangzhu gucheng yizhi taoqi de fenxi yanjiu 良渚古城遗址陶器的分析研究 (Analytical Studies on the Pottery Found in the Liangzhu Ancient City). *Zhongguo Kexue: Jishu Kexue 中国科学：技术科学 Scientia Sinica (Technologica)* 43: 460–466.

Lu, Y. H. 陆耀华 (1991). Zhejiang jiaxing dafen yizhi de qingli 浙江嘉兴大坟遗址的清理 (Excavation of the Dafen Site in Zhejiang). *Wenwu 文物 (Cultural Relics)* (7): 1–9.

Luan, F. S. (2013). The Dawenkou Culture in the Lower Yellow River and Huai River Basin Areas, in A. Underhill (ed.), *A Companion to Chinese Archaeology*. Hoboken: John Wiley & Sons, pp. 411–434.

Luo, J. J. 罗今见 (2001). Dui 'rending shengtian' de lishi fansi 对'人定胜天'的历史反思 (Historical Rethinking on the 'Man Can Conquer Nature'). *Ziran Bianzhengfa Tongxu 自然辩证法通讯 (Journal of Dialectic of Nature)* (5): 68–73.

Lv, B. W., et al. 吕不韦等 (1996). *Lvshi Chunqiu 吕氏春秋 (Lv's Annals)*. Shanghai: Shanghai Ancient Books.

Ma, X. L., et al. 马萧林等 (2006). Lingbao xipo yangshao wenhua mudi chutu yuqi chubu yanjiu 灵宝西坡仰韶文化墓地出土玉器初步研究 (Preliminary Research on the Jade Discovered from the Yangshao Culture Cemetery at Xipo, Lingbao). *Zhongyuan Wenwu 中原文物 (Cultural Relics of Central China)* (2): 69–74.

Mann, M. (2012). *The Sources of Social Power. Volume 1: A History of Power from the Beginning to AD 1760, 2nd Edition*. Cambridge: Cambridge University Press.

Matthews, R. (2003). The Archaeology of Mesopotamia: Theories and Approaches. New York: Routledge.

Ministry of Science and Technology of the People's Republic of China 中华人民共和国科技部 and State Administration of Cultural Heritage 国家文物局 (eds.). (2009). *Zhonghua Wenming Tanyuan Gongcheng Wenji: Jishu yu Jingji Juan 中华文明探源工程文集-技术与经济卷 (Work Collections for the Research*

on the Origin of Chinese Civilization: The Technology and Economics Volume).
Beijing: Science Press.

Mithen, S., Black, E. (eds.). (2011). *Water, Life and Civilisation: Climate, Environment and Society in the Jordan Valley.* Cambridge: Cambridge University Press.

Mou, Y. K. 牟永抗 (1989a). Liangzhu yuqi santi 良渚玉器三题 (Three Issues about the Liangzhu Jade). *Wenwu* 文物 *(Cultural Relics)* (5): 64–68.

———. (1989b). Liangzhu yuqi shang shenchongbai de tansuo 良渚玉器上神崇拜的探索 (Study on the Divine Worship of Liangzhu Jade), in The Editorial Board (ed.), *Qingzhu Su Binqi Kaogu Wushiwu Nian Lunwenji* 庆祝苏秉琦考古五十五年论文集 *(Essay Collection Commemorating the Fifty-Fifth Year of Su Binqi's Archeological Work).* Beijing: Cultural Relics Press, pp. 184–197.

———. (1997). Shilun yuqi shidai: Zhongguo wenming shidai chansheng de yige zhongyao biaozhi 试论玉器时代：中国文明时代产生的一个重要标志 (A Discussion of the Jade Age: An Important Hallmark of the Emergence of Civilization in China), in B. Q. Su 苏秉琦 (ed.), *Kaoguxue Wenhua Lunji* 考古学文化论集 *(Collected Essays on Archaeological Cultures), Vol. 4.* Beijing: Cultural Relics Press.

———. (2003). Guanyu shiqian zhuoyu gongyi kaoguxue yanjiu de yixie kanfa 关于史前琢玉工艺考古学研究的一些看法 (Some Insights on the Archaeological Research of Prehistoric Jade-Working Techniques), in X. H. Qian 钱宪和 and J. N. Fang, 方建能 (eds). *Shiqian Zhuoyu Gongyi Jishu* 史前琢玉工艺技术 *(Prehistoric Jade-Working Techniques).* Taipei: National Taiwan Museum, 20.

Nakamura, S. 中村慎一 (2003). Liangzhu wenhua de yizhiqun 良渚文化的遗址群 (Liangzhu Site Clusters). *Gudai Wenming* 古代文明 *(Ancient Civilizations)* (2): 53–64.

Nanjing Museum 南京博物院. (1984). 1982 nian jiangsu changzhou wujin sidun yizhi de fajue 1982 年江苏常州武进寺墩遗址的发掘 (Excavation of the Sidun Site in Wujin, Changzhou, Jiangsu, 1982). *Kaogu* 考古 *(Archaeology)* (2): 109–129.

———. (2001). Jiangsu gourong dingshadi yizhi dierci fajue jianbao 江苏句容丁沙地遗址第二次发掘简报 (Brief Report of the Second Excavation at the Dingshadi Site in Gurong, Jiangsu). *Wenwu* 文物 *(Cultural Relics)* (5): 22–36.

———. (2003). *Huating: Xinshiqi Shidai Mudi Fajue Baogao* 花厅：新石器时代墓地发掘报告 *(Huating: Excavation Report of the Neolithic Cemetery).* Beijing: Cultural Relics Press.

———. (2012). *Zhaolingshan: 1990–1995 Niandu Fajue Baogao* 赵陵山：1990–1995 年度发掘报告 *(Zhaolingshan: Excavation Reports of the 1990–1995 Seasons).* Beijing: Cultural Relics Press.

Nanjing Museum, et al. 南京博物院等 (2010). Jiangshu zhangjiagang shi dongshancun xinshiqi shidai yizhi 江苏张家港市东山村新石器时代遗址 (The Neolithic Site at Dongshan Village, Zhangjiagang, Jiangsu Province). *Wenwu* 文物 *(Cultural Relics)* (8): 3–12.

———. (2016). *Dongshancun* 东山村 *(The Dongshancun Site).* Beijing: Cultural Relics Press.

National Museum of China 中国国家博物馆 and Zhejiang Cultural Heritage Bureau 浙江文物局. (2005). *Wenming de Shuguang: Liangzhu Wenhua Wenwu Jingpinji* 文明的曙光：良渚文化文物精品集 *(The Dawn of Civilisation: Finest Objects from the Liangzhu Culture).* Beijing: Chinese Social Sciences Press.

Nelson, S. M. (2008). *Shamanism and the Origin of States: Spirits, Power, and Gender in East Asia.* Walnut Creek, CA: Left Coast.

Ningxia Provincial Institute of Cultural Relics and Archaeology 宁夏文物考古研究所 and Historical Museum of China 中国历史博物馆. (2003). *Ningxia Caiyuan: Xinshiqi Shidai Yizhi, Muzang Fajue Baogao* 宁夏菜园: 新石器时代遗址墓葬发掘报告 *(Ningxia Caiyuan: The Excavation Report of Neolithic Sits and Cemetery).* Beijing: Science Press.

Nissen, H. J. (2001). Cultural and Political Networks in the Ancient Near East during the Fourth and Third Millennia BC, in M. S. Rothman (ed.), *Uruk Mesopotamia and Its Neighbors: Cross-Cultural Interactions in the Era of State Formation.* Santa Fe, NM: School of American Research Press, pp. 149–180.

Nissen, H. J., Damerow, P., Englund, R. K. (1993). *Archaic Bookkeeping: Early Writing and Techniques of Economic Administration in the Ancient Near East.* Chicago: University of Chicago Press.

Okamura, H. 冈村秀典 (2012). Chūgoku saiko no kyūtei girei o suru 中国最古の宫廷儀礼をする (The Oldest Aulic Etiquette in China), in *NHK 'Mystery of Chinese Civilization' Interview Team, Chū natsu bunmei no tanjyou*中夏文明の誕生 *(The Birth of Xia Dynasty in China).* Tokyo: 講談社 (Kodansha), pp. 96–105.

Ortmann, A. L., Kidder, T.R. (2013). Building Mound A at Poverty Point, Louisiana: Monumental Public Architecture, Ritual Practice, and Implications for Hunter-Gatherer Complexity. *Geoarchaeology* 28: 66–86.

Peking University School of Archaeology and Museology 北京大学考古文博学院, Zhejiang Provincial Institute of Cultural Relics and Archaeology 浙江省文物考古研究所, and Sophia University じょうちだいがく. (1998). Zhejiang tongxiang puanqiao yizhi fajue *jianbao* 浙江桐乡普安桥遗址发掘简报 (Brief Excavation Report of the Pu'anqiao Site, Tongxiang, Zhejiang). *Wenwu*文物 *(Cultural Relics)* (4): 61–74 and 97.

Peterson, C. E., Lu, X. M. (2013). Understanding Hongshan Period Social Dynamics, in A. Underhill (ed.), *A Companion to Chinese Archaeology.* Hoboken, NJ: Wiley, pp. 55–80.

Pollock, S. (1999). *Ancient Mesopotamia.* Cambridge: Cambridge University Press.

———. (2001). The Uruk Period in Southern Mesopotamia, in M. S. Rothman (ed.), *Uruk Mesopotamia and Its Neighbors: Cross-Cultural Interactions in the Era of State Formation.* Santa Fe, NM: School of American Research Press, pp. 181–232.

Qin, L. 秦岭 (2005). Fuquanshan mudi yanjiu 福泉山墓地研究 Study of the Fuquanshan Cemetery. *Gudai Wenming 古代文明 (Ancient Civilizations)* 4: 1–36.

———. (2013). The Liangzhu Culture, in A. Underhill (ed.), *A Companion to Chinese Archaeology.* Hoboken, NM: Wiley, pp. 574–596.

———. (2014). Tongxiang pu'anqiao yizhi zaoqi muzang ji songze fengge yuqi 桐乡普安桥遗址早期墓葬及崧泽风格玉器 (The Early Phase Burials and Songze-Style Jade Objects from the Pu'anqiao Site, Tongxiang), in Zhejiang Provincial Institute of Cultural Relics and Archaeology 浙江省文物考古研究所 (ed.), *Zhebei Songze Wenhua Kaogu Baogaoji (1996–2014)* 浙北崧泽文化考古报告集 *(1996–2014) (Collection of Excavation Reports of Songze Culture Sites in Northern Zhejiang).* Beijing: Cultural Relics Press, pp. 134–152.

Qin, L. 秦岭, Cui, J. F. 崔剑锋, Yang, Y. L. 杨颖亮 (2015). Xiaodouli yizhi chutu yushiqi dechubu fenxi 小兜里遗址出土玉石器的初步分析 (The Scientific Analysis of Jades from the Xiaodouli site), in Zhejiang Provincial Institute of Cultural Relics and Archaeology 浙江省文物考古研究所, Haining Museum 海宁博物馆 (eds.), *Xiaodouli 小兜里 (Excavation Report of the Xiaodouli Site)*. Beijing: Cultural Relics Press, pp. 412–426.

Qiu, X. G. 裘錫圭 (1980). Jiaguwen zhong de jizhong yueqi mingcheng 甲骨文中的几种乐器名称 (The Name of a Few Instruments Recorded in Oracle Bone Inscriptions). *Zhonghua Wenshi Luncun 中华文史论丛 (Journal of Chinese Literature and History)* (2): 67–79.

Rawson, J. (1995). Chinese Jade from the Neolithic to the Qing. London: British Museum Press.

Renfrew, C., Bahn, P. (2004). *Archaeology: Theory, Methods and Practice*. London: Thames and Hudson.

Sassaman, K. E. (2005). Poverty Point as Structure, Event, Process. *Journal of Archaeological Method and Theory* 12: 335–364.

Shaanxi Provincial Institute of Archaeology 陕西省考古研究院 and Shangluo Museum 商洛市博物馆. (2011). *Shangluo Donglongshan 商洛东龙山*. Beijing: Science Press.

Shaanxi Provincial Institute of Archaeology, et al. 山西省考古研究所等 (2002). Shanxi ruicheng qingliangsi mudi yuqi 山西芮城清凉寺墓地玉器 (Jade Objects from the Qingliangsi Cemetery in Ruicheng, Shanxi). *Kaogu yu Wenwu 考古与文物 (Archaeology and Cultural Relics)* (5): 3–7.

———. (2006). Shanxi ruicheng qingliangsi xinshiqi shidai mudi 山西芮城清凉寺新石器时代墓地 (Neolithic Cemetery at Qingliangsi, Ruicheng, Shanxi). *Wenwu 文物 (Cultural Relics)* (3): 4–17.

———. (2011). Shanxi ruicheng qingliangsi shiqian mudi 山西芮城清凉寺史前墓地 (Prehistoric Cemetery at Qingliangsi, Ruicheng, Shanxi). *Kaogu Xuebao 考古学报 (Acta Archaeologica Sinica)* (4): 525–573.

———. (2013). Shanxi Shenmuxian Shimao Yizhi 陕西神木县石峁遗址 (Shimao Site in Shenmu County, Shaanxi). *Kaogu 考古 (Archaeology)* (7): 14–24.

Shanghai Committee of Cultural Relics Management 上海文物管理委员会. (2000). *Fuquanshan: Xinshiqi Shidai Yizhi Fajue Baogao 福泉山: 新石器时代遗址发掘报告 (Fuquanshan: Excavation Report of a Neolithic Site)*. Beijing: Cultural Relics Press.

Shanghai Museum 上海博物馆. (2014). *Shencheng Xunzong: Shanghai Kaogu Dazhan 申城寻踪: 上海考古大展 (Tracing History: Exhibition of Archaeological Discoveries in Shanghai)*. Shanghai: Shanghai Book and Painting Press.

———. (2015). Shanghai fuquanshan yizhi wujiachang mudi 2010 nian fajue jianbao 上海福泉山遗址吴家场墓地2010年发掘简报 (The 2010 Excavation Report of the Wujiaochang Cemetery at the Shanghai Fuquanshan Site). *Kaogu 考古 (Archaeology)* (10): 46–65.

Shelach, G., Teng, M. Y. (2013). Earlier Neolithic Economic and Social Systems of the Liao River Region, Northeast China, in A. Underhill (ed.), *A Companion to Chinese Archaeology*. Hoboken, NJ: Wiley, pp. 35–54.

Shen, C. Y. 沈长云, Li, J. 李晶 (2004). Chunqiu guanzhi yu zhouli bijiao yanjiu: Zhouli chengshu niandai zaitan 春秋官制与《周礼》比较研究-《周礼》成书年代再探讨 (A Comparative Study of the Bureaucratic System in the Spring and Autumn Period and That in the Zhou Li: Further Discussion of the Compilation Date of Zhou Li). *Lishi Yanjiu 历史研究 (Historical Research)* (6): 3–26.

Sheng, Q. 盛启新 (2014). Songze Liangzhu Wenhua de Wenshi Xitong 崧泽、良渚文化的纹饰系统 (The Motif Systems of the Songze and Liangzhu

Cultures). Doctoral Thesis for the School of Archaeology and Museology, Peking University.

Shi, X. G. 施昕更 (1938). *Liangzhu (Hangxian Dierqu Heitao Wenhua Yizhi Chubu Baogao)* 良渚(杭县第二区黑陶文化遗址初步报告) *(Liangzhu: A Preliminary Report on the Black Pottery Culture Site in the Second District of the Hang County)*. Hangzhou: Zhejiang Provincial Department of Education.

Sichuan Provincial Institute of Cultural Relics and Archaeology 四川省文物考古研究所. (1999). *Sanxingdui Jisikeng* 三星堆祭祀坑 *(Sacrificial Pits at the Sanxingdui Site)*. Beijing: Cultural Relics Press.

Sima, Q. 司马迁. (1982). *Shi Ji* 史记 *(Records of the Grand Historian)*. Beijing: Zhonghua Book.

Song, J. 宋建 (2000). Guanyu songze wenhua zhi liangzhu wenhua guodu jieduan de jige wenti 关于松泽文化至良渚文化过渡阶段的几个问题 (Questions on the Transition Period between the Songze Culture and the Liangzhu Culture). *Kaogu* 考古 *(Archaeology)* (11).

———. (2014). Zhongguo zaoqi wenming Jincheng zhong de guguo: Lingjiatan he fuquanshan 中国早期文明进程中的古国: 凌家滩和福泉山 (Lingjiatan and Fuquanshan: Archaic States in Early Chinese Civilisation), in Shanghai Museum 上海博物馆 (ed.). *'Chengshi yu Wenming' Guoji Xueshu Yantaohui Lunwengao* '城市与文明' 国际学术研讨会论文稿 *(Papers from the International Conference 'Cities and Civilisations')*. Shanghai: Shanghai Ancient Books, pp. 232–248.

Song, J. Z. 宋建忠 (2003). Shanxi linfen xiajin mudi yushiqi fenxi 山西临汾下靳墓地玉石器分析 (Analysis of Jade and Stone Objects from Xiajin Cemetery in Linfen, Shanxi). *Gudai Wenming* 古代文明 *(Ancient Civilisations)* 2: 121–137.

Spengler, R., et al. (2014). Early Agriculture and Crop Transmission among Bronze Age Mobile Pastoralists of Central Eurasia. *Proceedings of the Royal Society of London B: Biological Sciences* 281: 20133382.

Su, B. Q. 苏秉琦 (2000). *Zhongguo Wenming Qiyuan Xintan* 中国文明起源新探 *(A New Exploration of the Origin of Chinese Civilization)*. Beijing: Sanlian Publishing.

Sun, B. (2013). The Longshan Culture of Shandong, in A. Underhill (ed.), *A Companion to Chinese Archaeology*. Hoboken, NJ: Wiley, pp. 435–458.

Sun, Q. W. 孙庆伟 (2000). 'Kaogongji: yuren' de kaoguxue yanjiu 《考工记·玉人》的考古学研究 (Examining *Record of Trades: Jade Figurine* from an Archaeological Perspective), in School of Archaeology and Museology Peking University (ed.), *Kaoguxue Yanjiu* 考古学研究 *(Archaeological Research)*, Vol. 4. Beijing: Science Press, pp. 115–139.

———. (2002). 'Zuozhuan' suojian yongyu shili yanjiu 《左传》所见用玉事例研究 (Examples of Jade Use in *Zuo Zhuan*). *Gudai Wenming* 古代文明 *(Ancient Civilisations)* 1: 310–370.

———. (2003). Zhoudai jisi ji qi yongyu santi 周代祭祀及其用玉制度三题 (Three Issues on Rituals and Aystematic Jade Use during the Zhou Dynasty). *Gudai Wenming* 古代文明 *(Ancient Civilisations)* 2: 213–229.

———.(2004). Zhoudaijinwensuojianyongyushiliyanjiu 周代金文所见用玉事例研究 (Examples of Jade Use Recorded on Zhou Dynasty Bronze Inscriptions). *Gudai Wenming* 古代文明 *(Ancient Civilisations)* 3: 320–342.

———. (2008). *Zhoudai yongyu zhidu yanjiu* 周代用玉制度研究 *(System of Jade Use during the Zhou Dynasty)*. Shanghai: Shanghai Ancient Books.

————. (2012). Shangyu mingwen yu zhoudai de pinli 尚盂铭文与周代的聘礼 (The Shangyu Inscription and Diplomatic Exchange during the Zhou Dynasty), in School of Archaeology and Museology Peking University (ed.), *Kaoguxue Yanjiu 考古学研究 (Archaeological Research), Vol. 10.* Beijing: Science Press.

————. (2013). Lishi Qiuzhuye: Shilun 'Yazhang' de Yuanliu yu Mingcheng 礼失求诸野：试论'牙璋'的源流与名称 (Searching for Lost Rites Outside the Main Court: A Discussion About the Origin and Name of the *Yazhang* Tablet). Yuqi Kaogu Tongxun 玉器考古通讯 (*Correspondence of Archaeology of Jade*) (2): 467–508.

Thorp, R. L. (2006). *China in the Early Bronze Age: Shang Civilization.* Philadelphia: University of Pennsylvania Press.

Voegelin, E. (2010). *Yiselie yu Qishi* 以色列与启示*Israel and Revelation.* W. A. Huo 霍伟岸 and Y. Ye 叶颖 (trans.). Nanjing: Yilin Press.

von Falkenhausen, L. (2006). *Chinese Society in the Age of Confucius (1000–250 BC): The Archaeological Evidence.* Los Angeles: Cotsen Institute of Archaeology.

Wagner, M., et al. (2013). Mapping of the Spatial and Temporal Distribution of Archaeological Sites of Northern China during the Neolithic and Bronze Age. *Quaternary International* 290: 344–357.

Wang, G. W. 王国维 (1959). *Guantang Jilin* 观堂集林 *(The Work Collection of Guan-tang).* Beijing: Zhonghua Book.

Wang, M. D. 王明达 (1989). Fanshan liangzhu wenhua mudi chulun反山良渚文化墓地初论 (Preliminary Thoughts on the Liangzhu Cemetery at Fanshan). *Wenwu 文物 (Cultural Relics)* (10): 48–52.

————. (2003). Jieshao yijian liangzhu wenhua yucong banchengpin: Jiantan cong de zhizuo gongyi 介绍一件良渚文化玉琮半成品：兼谈琮的制作工艺 (Introducing an Unfinished Liangzhu Jade Cong Tube and Discussing Techniques Used to Produce Cong Tubes), in X. H. Qian 钱宪和, J. N. Fang 方建能 (eds). *Shiqian Zhuoyu Gongyi Jishu 史前琢玉工艺技术 (Prehistoric Jade-Working Techniques).* Taipei: National Taiwan Museum, pp. 607–611.

————. (2006). Liangzhu yizhiquan zai renshi良渚遗址群再认识 (A Reappraisal of the Liangzhu Site Cluster), in Zhejiang Provincial Institute of Archaeology and Cultural Relics 浙江省文物考古研究所 (ed.), *Zhejiangsheng Wenwu Kaogu Yanjiusuo Xuekan 浙江省文物考古研究所学刊(Academic Journal of the Zhejiang Provincial Institute of Cultural Relics and Archaeology),* Vol. 8. Beijing: Science Press, pp. 393–400.

Wang, N. Y. 王宁远 (2009). Liangzhu yizhiqun houyangcun yizhi 良渚遗址群后杨村遗址 (Houyangcun Site of the Liangzhu Site Cluster), in Zhejiang Provincial Institute of Cultural Relics and Archaeology 浙江省文物考古研究所 (ed.), *Zhejiang Kaogu Xinjiyuan 浙江考古新纪元 (A New Era of Archaeology in Zhejiang).* Beijing: Science Press, pp. 131–132.

————. (2013). *Cong Cunju dao Wangcheng* 从村居到王城 *(From Village Settlement to Royal City).* Hangzhou: Hangzhou Press.

Wang, N. Y. 王宁远, Yan, K. K. 闫凯凯. (2014). Liangzhu Xianmin de Zhishui Shijian yu Shanggu Zhishui Chuanshuo 良渚先民的治水实践与上古治水传说 (Water Management Practices of the Liangzhu People and Ancient Legends on Water Management), in *Yuhuicun Yizhi yu Huaihe Liuyu Wenming Yantaohui Lunwenji 禹会村遗址与淮河流域文明研讨会论文集 (Collected Essays from the Conference on the Yuhuicun Site and the Civilisation of the Huai River Region).* Beijing: Cultural Relics Press, pp. 195–204.

Wang, N. Y., Dong, C. W., Xu, H. G. (2019). *Liangzhu Gucheng Chengqiang Pudianshi Laiyuan ji Gongcheng Yanjiu Baogao* 良渚古城城墙铺垫石来源及工程研究报告 *(Provenance and Engineering of the Liangzhu City Rock Base)*. Hangzhou: Zhejiang Ancient Books.

Wang, N. Y., et al. 王宁远等 (2009). Tongxiang yaojiashan 桐乡姚家山, in Zhejiang Provincial Institute of Cultural Relics and Archaeology 浙江省文物考古研究所 (ed.), *Zhejiang Kaogu Xinjiyuan* 浙江考古新纪元 *(A New Era of Archaeology in Zhejiang)*. Beijing: Science Press, pp. 88–91.

Wang, R. B. 汪荣宝 (1996). Fanyan Yishu 法言義疏 (Interpretation of Examplry Sayings by Yang Xiong). Beijing: Zhonghua Book.

Wang, S. S. 王世舜, Wang, C. Y. 王翠叶 (2011). *Shangshu* 尚书 *(Classic of History.* Beijing: Zhonghua Book.

Wang, Z. 汪遵国 (1984). Liangzhu Wenhua 'Yulianzang' Lueshu 良渚文化 '玉敛葬' 略述 (Brief Discussion of the 'Jade Burial' in the Liangzhu Culture). *Wenwu* 文物 *(Cultural Relics)* (2): 23–26, 100–101.

Wen, G. 闻广 (1986). Sunan xinshiqi shidai yuqi de kaogu dizhixue yanjiu 苏南新石器时代玉器的考古地质学研究 (Archaeological and Geological Analysis of Neolithic Jade from Southern Jiangsu). *Wenwu* 文物 *(Cultural Relics)* (10): 42–49.

———. (1993). Yu yu min 玉与珉 (Jade and Pseudo-Jade). *Gugong Wenwu Yuekan* 故宫文物月刊 *(Cultural Relics of Palace Museum)* 124(4): 126–137.

———. (1994). Guyu de shouqin: Guyu congtan liu 古玉的受沁—古玉丛谈六 (Post-Depositional Changes of Ancient Jade: Collected Essays on Ancient Jade, Part Six). *Gugong Wenwu Yuekan* 故宫文物月刊 *(Cultural Relics of Palace Museum)* (12): 92–101.

———. (1995). Guyu de xueqin 古玉的血沁 ('Xueqin' Stage of Post-Depositional Changes of Ancient Jade), *Gugong Wenwu Yuekan* 故宫文物月刊 *(Cultural Relics of Palace Museum)* 133(1): 92–101.

Wen, G. 闻广, Jing, Z. C. 荆志淳 (2000). Zhongguo guyu dizhi kaoguxue yanjiu: Fuquanshan yuqi, fu Songze yuqi 中国古玉地质考古学研究：福泉山玉器，附崧泽玉器 (Archaeological and Geological Study of Ancient Chinese Jade: Jade Objects from Fuquanshan and Songze), in Shanghai Committee of Cultural Relics Management 上海文物管理委员会, *Fuquanshan: Xinshiqi Shidai Yizhi Fajue Baogao* 福泉山：新石器时代遗址发掘报告 *(Fuquanshan: Excavation Report of a Neolithic Site)*. Beijing: Cultural Relics Press, pp. 193–216.

Whitehead, P. G., et al. (2008). Modelling of Hydrology and Potential Population Levels at Bronze Age Jawa, Northern Jordan: A Monte Carlo Approach to Cope with Uncertainty. *Journal of Archaeological Science* 35: 517–529.

Wu, R. Z 吴汝祚 (1990). Shitan yuqi shidai: Zhonghua wenming qiyuan de tansuo 试探玉器时代：中华文明起源的探索 (A Discussion of the Jade Age: Exploring the Origin of Chinese Civilisation). *Zhongguo Wenwubao* 中国文物报 *(China Cultural Relics News)*, 1 November.

Xu, H. 许宏 (2009). *Zuizao De Zhongguo* 最早的中国 *(The Earliest China)*. Beijing: Science Press.

Xu, S. 许慎 (1963). Yu Bu 玉部 (Jade Section), in *Shuowen Jieji* 说文解字 *(Explaining Graphs and Analyzing Characters)*. Beijing: Zhonghua Book.

Xu, Y. H., et al. 徐元诰等 (2002). *Guoyu Jijie* 国语集解 *(The Collection of Comments on the Discourses of the States)*. Beijing: Zhonghua Book.

Yan, W. M. 严文明 (1996). Liangzhu Suibi 良渚随笔 (Essay on Liangzhu). *Wenwu* 文物 (3): 319–329.

——. (1997). Huanghe liuyu wenming de faxiang yu fazhan 黄河流域文明的发祥与发展 (The Birth and Development of the Yellow River Civilization). *Huaxia Kaogu* 华夏考古 *(Huaxia Archaeology)* (1): 83–89.

Yan, Y. L. 闫亚林. (2010). Xibei Diqu Shiqian Yuqi Yanjiu 西北地区史前玉器研究 (Research on the Neolithic Jade Objects from Northwest China). Doctoral dissertation, School of Archaeology and Museology, Peking University.

Yang, B. 杨波 (1996). Shandong wulianxian dantu yizhi chutu de yuqi 山东五莲县丹土遗址出土的玉器 (Jade Objects Excavated at the Dantu Site in Wulian County, Shandong). *Gugong Wenwu Yuekan* 故宫文物月刊 *(Cultural Relics of Palace Museum)* (158): 84–95.

Yang, B. D. 杨伯达 (2011). Preface, in *Wuyu zhi guang: Xuji* 巫玉之光-续集 *(The Glory of Wu Jade: A Sequel)*. Beijing: Forbidden City Publishing, pp. 1–5.

Yang, B. J. 杨伯峻 (2006). *Weizheng* 为政, in *Lunyu Yizhu* 语译注 *(The Interpretation of Analects of Confucius)*. Beijing: Zhonghua Book, pp. 11–23.

Yang, J. H. 杨建华 (2014). *Lianghe Liuyu: Cong Nongye Cunluo Zouxiang Chengbang Guojia* 两河流域: 从农业村落走向城邦国家 *(Mesopotamia: From Agricultural Villages to City States)*. Beijing: Science Press.

Yanshi County Museum 偃师县文化馆. (1978). Erlitou yizhi chutu de tongqi he yuqi 二里头遗址出土的铜器和玉器 (Bronze and Jade Excavated at the Erlitou Site). *Kaogu* 考古 *(Archaeology)* (4): 270–300.

Ye, M. Z. 叶美珍 (2005). Beinan yizhi zhi yuqi wenhua 卑南遗址之玉器文化 (Jade Culture at the Beinan Site), in Z. H. Zang, 臧振华, M. Z. Ye 叶美珍 (eds.), *Guancang Beinan Yizhi Yuqi Tulu* 馆藏卑南遗址玉器图录 *(Illustrated Catalogue of Jade from Beinan Site in National Museum of Prehistory)*. Taipei: National Museum of Prehistory.

Yuhang Committee of Heritage Management 浙江省余杭县文管会. (1996). Zhejiang yuhang hengshan liangzhu wenhua muzang qingli jianbao 浙江余杭横山良渚文化墓葬清理简报 (A Brief Excavation Report of the Liangzhu Burials at Hengshan, Yuhang, Zhejiang), in Nanjing Museum (ed.), *Dongfang Wenming Zhi Guang: Liangzhu Wenhua Faxian 60 Zhounian Jinian Wenji* 东方文明之光: 良渚文化发现60周年纪念文集 *(The Dawn of Chinese Civilisation: Collected Articles Commemorating the 60th Anniversary of the Discovery of the Liangzhu Culture)*. Hainan: Hainan Guojixinwen, pp. 69–77.

Zang, Z. H. 藏振华 (2005). Haiyu qipa: Taiwan dong'an de yuqi wenming 海隅奇葩-台湾东岸的玉器文明(Pearl of the Sea Corner: Jade Civilisation from East Taiwan), in Z. H. Zang, M. Z. Ye 叶美珍 (eds.), *Guancang Beinan Yizhi Yuqi Tulul* 馆藏卑南遗址玉器图录 *(Catalogue of Jade from the Beinan Site)*. Taipei: Museum of Prehistoric Culture of Taiwan, pp. 3-16.

Zhang, C. 张弛 (1997). Liangzhu wenhua damu shixi 良渚文化大墓试析 (Analysis on the Large Tombs of the Liangzhu Culture) in School of Archaeology and Museology Peking University (ed.), *Kaoguxue Yanjiu* 考古学研究 *(Archaeological Research)*. Beijing: Science Press, pp. 57–67.

——. (2013). The Qujialing–Shijiahe Culture in the Middle Yangzi River Valley, in A. Underhill (ed.), *A Companion to Chinese Archaeology*. Hoboken, NJ: Wiley, pp. 510–534.

——. (2017). Longshan-erlitou: Shiqian wenhua geju de gaibian yu qingtong shidai quanqiuhua de xingcheng 龙山-二里头: 中国史前文化格局的改变与青铜时代全球化的形成 (Longshan – Erlitou: The Change of Chinese Prehistoric Culture and the Formation of Globalisation in the Bronze Age. *Wenwu* 文物 *(Cultural Relics)* (6): 50–59.

Zhang, G. Z. 张光直 (1986). Tan 'Cong' jiqi zai zhongguo gushi shang de yiyi 谈'琮'及其在中国古史上的意义 (A Discussion on 'Cong' and Its Significance in Chinese History), in Wenwu Chubanshe Bianjibu 文物出版社编辑部 (The Editorial Department of the Cultural Relics Press) (ed.), *Wenwu yu Kaogu Lunji* 文物与考古论集 *(Collection of Archaeology and Cultural Relic Studies)*. Beijing: Culture Relics Press, pp. 252–260.

———. (1988). Puyang sanqiao yu zhongguo gudai meishu shang de renshou muti 濮阳三蹻与中国古代美术上的人兽母题 (The Puyang Three Qiao and the Human and Beast Motif in Chinese Ancient Art). *Wenwu* 文物 *(Cultural Relics)* (11): 36–39.

Zhang, J. (2006). Lingjiatan yuren de yishu jiedu 凌家滩玉人的艺术解读 (An Artistic Interpretation of the Jade Figures from Lingjiatan), in Anhui Provincial Institute of Cultural Relics and Archaeology (ed.), *Lingjiatan Wenhua Yanjiu* 凌家滩文化研究 *(Studies on the Lingjiantan Culture)*. Beijing: Cultural Relics Press, pp. 181–185.

Zhang, J. G., et al. 张敬国等 (2006a). Pianzhuang gongju kailiao zhi chubu shiyan: Yuqi diaozhuo gongyi xianwei tansuo zhi san 片状工具开料之初步试验: 玉器雕琢工艺显微探索之三 (Preliminary Experiment Using Blade Tools in Jade Cutting: Microscopic Examination of Jade-Working Techniques, Part III). *Yuwenhua Luncong* 玉文化论丛 *(Forum of Jade Cultures)* (1): 311–326.

———. (2006b). Xianxing Gongju Kailiao zhi Chubushiyan: Yuqi Diaozhuo Gongyi Xianwei Tansuo Zhiyi 线性工具开料之初步试验: 玉器雕琢工艺显微探索之一 (Preliminary Experiment on Using String Tools in the Initial Steps of Jade-Working: Microscopic Exploration of Jade-Working Techniques, Part I). *Yuwenhua Luncong* 玉文化论丛 *(Forum of Jade Cultures)* (1): 295–303.

Zhang, M. 张敏 (2013). Zhiyushuo 治玉说 (On the Jade-Working Techniques), in Nanjing Museum (ed.), *Zhang Min Wenji* 张敏文集 *(Collected Works of Zhang Min)*. Beijing: Cultural Relics Press, pp. 195–202.

Zhang, X. 张辛 (2003). Yuqi liyi lunyao 玉器礼义论要 (Discussing the Key Meaning and Rituals of Jade Objects). *Zhongguo Lishi Wenwu* 中国历史文物 *(Journal of National Museum of China)* (6): 28–37.

Zhang, X. Q. 张绪球 (2008). *Shijiahe Wenhua Yuqi* 石家河文化玉器 *(Jade Objects from the Shijiahe Culture)*. Beijing: Cultural Relics Press.

Zhang, Z. P. 张忠培 (1995). Liangzhu wenhua de niandai he qi suochu shehui jieduan: wuqian nianqian zhongguo jinru wenming de yige lizheng 良渚文化的年代和其所处社会阶段: 五千年前中国进入文明的一个例证 (The Chronology and Social Development of the Liangzhu Culture: An Illustration of Chinese Civilization 5,000 Years Ago). *Wenwu* 文物 *(Cultural Relics)* (5): 47–58.

———. (2012). Liangzhu wenhua mudi yuqi biaoshu de wenming shehui 良渚文化墓地与其表述的文明社会 (The Liangzhu Culture Cemeteries and the Stratified Society They Represented). *Kaogu Xuebao* 考古学报 *Acta Archaeologica Sinica* (4): 401–422.

Zhao, C. Q. 赵春青 (2013). The Longshan Culture in Central Henan Province, c. 2600–1900 BC, in A. Underhill (ed.), *A Companion to Chinese Archaeology*, Hoboken, NJ: John Wiley & Sons, pp. 236–254.

Zhao, H. 赵辉 (1999). Liangzhu wenhua de ruogan teshuxing: Lun yichu zhongguo shiqian wenming de shuailuo yuanyin 良渚文化的若干特殊性-论一处中国史前文明的衰落原因 (The Specificities of the Liangzhu Culture: The Reasons behind the Decline of a Chinese Prehistoric Civilization), in Zhejiang Provincial Institute of Cultural Relics and Archaeology (ed.), *Liangzhu Wenhua*

Yanjiu: Jinian Liangzhu Wenhua Faxian Liushi Zhounian Guoji Xueshu Taolunhui Wenji 良渚文化研究- 纪念良渚文化发现六十周年国际学术讨论会文集 *(Studies of the Liangzhu Culture: Proceedings of the International Symposium on the 60th Anniversary for the Discovery of Liangzhu Culture)*. Beijing: Science Press, pp. 104–119.

——. (2000). Zhongguo wenming qiyuan yanjiu zhong de yige jiben wenti 中国文明起源研究中的一个基本问题 (A Basic Question in Research of the Origin of Chinese Civilization), in W. M. Yan, Y. Yasuda (eds.), *Daozuo, Taoqi he Dushi de Qiyuan* 稻作、陶器和都市的起源 (The Origin of Rice, Pottery and Cities). Beijing: Cultural Relics Press: 135–142.

——. (2003). Kaoguxue guanyu zhongguo wenming qiyuan wenti de yanjiu 考古学关于中国文明起源问题的研究 (Archaeological Studies on the Origin of Chinese Civilization). *Gudai Wenming* 古代文明 *(Ancient Civilizations)* 2: 1–12.

——. (2017). Liangzhu de guojia xingtai 良渚的国家形态 (The State Pattern of Liangzhu). *Zhongguo Wenhua Yichan* 中国文化遗产 *(Cultural Heritage of China)*, (3): 22–28.

Zhao, H. 赵辉, Wei, J. 魏峻 (2002). Zhongguo xinshiqi shidai gudai chengzhi de faxian yu yanjiu 中国新石器时代古代城址的发现与研究 (Discoveries and Research of Prehistoric Walled Sites in China). *Gudai Wenming* 古代文明 *(Ancient Civilizations)* 1: 1–34.

Zhao, Y. 赵晔 (2012). Zhejiang yuhang linping yizhiqun de juluo kaocha 浙江余杭临平遗址群的聚落考察 (Investigation on the Settlement Pattern of the Linping Site Cluster, Yuhang, Zhejiang Province). *Dongnan wenhua* 东南文化 *(Southeast Culture)* (3): 31–39.

——. (2013). Guanjingtou: Daxiongshan qiuling shiqian wenhua de yige chuangkou 官井头: 大雄山丘陵史前文化的一个窗口 (Guanjingtou: A Window for the Prehistoric Culture of the Daxiong Mountains). *Dongfang Bowu* 东方博物 *(Oriental Cultural Relics)* (48): 5–19.

——. (2014a). Liangzhu yuqi wenshi xinzheng: Guanjingtou jijian xinying liangzhu yuqi de jiedu 良渚玉器纹饰新证-官井头几件新颖良渚玉器的解读 (New Evidence for the Liangzhu Jade Motifs: Interpreting the New Types of Jade Objects from the Guanjingtou Site), in Yuhun Guopo: Zhongguo Gudai Yuqi yu Chuantong Wenhua Xueshu Taolunhui Wenji (Vol. 6) 玉魂国魄—中国古代玉器与传统文化学术讨论会文集 *(The Spirit of Jade and the Soul of the Nation*: Corpus of the Symposium on Ancient Chinese Jade and Traditional Culture: Vol. 6). Zhejiang: Ancient Books Publishing, pp. 251–263.

——. (2014b). Zhejiang liangzhu guanjingtou yizhi 浙江良渚官井头遗址 (The Guanjingtou Site in Liangzhu, Zhejiang), State Administration of Cultural Heritage 国家文物局 (ed.), *2013 Zhongguo Zhongyao Kaogu Faxian* 2013 中国重要考古发现 *(Important Archaeological Finds in China in 2013)*. Beijing: Cultural Relics Press, p. 8.

Zhejiang Cultural Heritage Bureau 浙江省文物局. (2011). *Faxian Lishi: Zhejiang Xinshiji Kaogu Chengguo Zhan* 发现历史: 浙江新世纪考古成果展 *(Discovering History: Exhibition on Archaeological Works in the New Century in Zhejiang)*. Beijing: China Photographing Press.

Zhejiang Provincial Institute of Cultural Relics and Archaeology 浙江省文物考古研究所. (2001). Yuhang mojiaoshan yizhi 1992–1993 nian de fajue 余杭莫角山遗址1992–1993年的发掘 (1992–1993 Excavation of the Mojiaoshan Site at Yuhang. *Wenwu* 文物 *(Cultural Relics)* (12): 4–19.

———. (2002a). Zhejiang yuhang boyishan yizhi fajue jianbao 浙江余杭钵衣山遗址发掘简报 (Brief Excavation Report of the Boyishan Site, Yuhang, Zhejiang). *Wenwu* 文物 *(Cultural Relics)* (10): 67–75.

———. (2002b). Zhejiang yuhang shangkoushan yizhi fajue jianbao 浙江余杭上口山遗址发掘简报 (Brief Excavation Report of the Shangkoushan Site, Yuhang, Zhejiang). *Wenwu* 文物 *(Cultural Relics)* (10): 57–66.

———. (2003). *Yaoshan* 瑶山. Beijing: Cultural Relics Press.

———. (2005a). *Fanshan* 反山. Beijing: Cultural Relics Press.

———. (2005b). *Nanhebang: Songze Wenhua Yizhi Fajue Baogao* 南河浜-崧泽文化遗址发掘报告*(Nanhebang: Excavation Report of the Songze Culture Site)*. Beijing: Cultural Relics Press.

———. (2005c). *Liangzhu Yizhi Qun* 良渚遗址群 *(The Liangzhu Site Cluster)*. Beijing: Cultural Relics Press.

———. (2005d). *Miaoqian* 庙前. Beijing: Cultural Relics Press.

———. (2006a). *Xindili* 新地里. Beijing: Cultural Relics Press.

———. (2006b). *Pishan* 毘山. Beijing: Cultural Relics Press.

———. (2008). Hangzhou shi yuyang qu liangzhu gucheng yizhi 2006–2007 nian fajue 杭州市余杭区良渚古城遗址2006–2007年发掘 (The Excavations for the Ancient Liangzhu City in Yuhang, Hangzhou during 2006–2007). *Kaogu* 考古 *(Archaeology)* (7): 3–10 and plates 1–2.

———. (2011). *Wenjiashan* 文家山. Beijing: Cultural Relics Press.

———. (2014). *Bianjiashan* 卞家山. Beijing: Cultural Relics Press.

———. (2015a). Meirendi he biandanshan de fajue yu liangzhu gucheng waiguo de tansuo 美人地和扁担山的发掘与良渚古城外郭的探索 (Excavation of Meirendi and Biandanshan and Survey of the Outer City of the Liangzhu Centre). *Kaogu* 考古 *(Archaeology)* (1): 14–29.

———. (2015b). Hangzhou Shi Liangzhu Gucheng Waiwei Shuili Xitong de Diaocha 杭州市良渚古城外围水利系统的调查 (Survey of the Hydraulic System Outside the Walled City of Liangzhu, Hangzhou). *Kaogu* 考古 *(Archaeology)* (1): 3–14.

Zhejiang Provincial Institute of Cultural Relics and Archaeology Fanshan Archaeological Team 浙江省文物考古研究所反山考古队. (1988). Zhejiang yuhang fanshan liangzhu mudi fajue jianbao 浙江余杭反山良渚墓地发掘简报 (Brief Excavation Report of the Liangzhu Cemetery at Fanshan, Yuhang, Zhejiang). *Wenwu* 文物 *(Cultural Relics)* (1): 1–31.

Zhejiang Provincial Institute of Cultural Relics and Archaeology 浙江省文物考古研究所 and Liangzhu Museum 良渚博物馆. (2014). *Songze zhi Mei: Zhejiang Songze Wenhua Kaogu Tezhan* 崧泽之美: 浙江崧泽文化考古特展 *(The Beauty of Songze: Special Exhibition on the Archaeology of the Songze Culture in Zhejiang)*. Hangzhou: Zhejiang Photographing Press.

Zhejiang Provincial Institute of Cultural Relics and Archaeology 浙江省文物考古研究所 and Ningbo Municipal Institute of Cultural Relics and Archaeology 宁波市文化保护与考古研究所. (1993). Ningbo cihu yizhi fajue jianbao 宁波慈湖遗址发掘简报 (Brief Excavation Report of the Cihu Site in Ningbo). *Zhejiangsheng Wenwu Kaogu Yanjiusuo Xuekan* 浙江省文物考古研究所学刊 *(Academic Periodical of the Zhejiang Provincial Institute of Cultural Relics and Archaeology)*, 108.

Zhejiang Provincial Institute of Cultural Relics and Archaeology 浙江省文物考古研究所, Shanghai Heritage Management Committee 上海市文物管理委员会, and Nanjing Museum 南京博物馆. (1990). *Liangzhu Wenhua Yuqi* 良渚文化玉器 *(Jade of the Liangzhu Culture)*. Beijing: Cultural Relics Press.

Zhejiang Provincial Institute of Cultural Relics and Archaeology 浙江省文物考古研究所 and Xiaoshan Museum 萧山博物馆. (2004). *Kuahuqiao 跨湖桥*. Beijing: Cultural Relics Press.

Zhejiang Provincial Institute of Cultural Relics and Archaeology 浙江省文物考古研究所 and Yuhang Cultural Relics Management Committee 余杭文管会. (1997). Zhejiang yuhang huiguanshan liangzhu wenhua jitan yu mudi fajue jianbao 浙江余杭汇观山良渚文化祭坛与墓地发掘简报 (Brief Report of the Excavation of the Liangzhu Altar and Cemetery at Huiguanshan, Yuhang, Zhejiang). *Wenwu 文物 (Cultural Relics)* (7): 74–93.

———. (2001). Huiguanshan yizhi dierci fajue jianbao 汇观山遗址第二次发掘简报 (Preliminary Report of the Second Excavation at the Huiguanshan Site). *Wenwu 文物 (Cultural Relics)* (12): 36–40.

Zhejiang Provincial Institute of Cultural Relics and Archaeology 浙江省文物考古研究所, Yuyao Institute of Heritage Conservation and Management 余姚市文物保护管理所, and Hemudu Site Museum 河姆渡遗址博物馆. (2007). Zhejiang yuyao tianluoshan xinshiqi shidai yizhi 2004 nian fajue jianbao浙江余姚田螺山新石器时代遗址2004年发掘简报 (Brief Excavation Report of Season 2004 of the Neolithic site at Tianluoshan, Yuyao, Zhejiang). *Wenwu 文物 (Cultural Relics)* (11): 4–24.

Zheng , J. X. 郑杰祥 (1987). Shi li yu 释礼、玉 (Explaining the Characters *Li* and *Yu*), in C. W. Tian田昌五 (ed.), *Huaxia Wenming 华夏文明 (Chinese Civilisation)*, Vol. 1. Beijing: Peking University Press, pp. 355–367.

Zheng, X., et al. 郑玄等 (2010). *Zhouli Zhushu周礼注疏 (The Notes and Commentaries of the Rites of Zhou)*. Shanghai: Shanghai Ancient Books.

Zheng, Y. F. 郑云飞, Chen, X. G. 陈旭高, Ding, P. 丁品 (2014). Zhejiang yuhang maoshan yizhi gudaotian gengzuo yiji yanjiu 浙江余杭茅山遗址古稻田耕作遗迹研究 (Study of the Ancient Paddy Fields at the Maoshan Site, Yuhang, Zhejiang Province). *Quaternary Sciences* 34: 85–96.

Zhou, L. J. 周丽娟 (2010). Shanghai qingpu fuquanshan yizhi wujiachang didian kaogu fajue上海青浦福泉山遗址吴家场地点考古发掘 (Archaeological Excavation of the Wujiachang Location at the Fuquanshan Site, Qingpu, Shanghai), in State Administration of Cultural Heritage 国家文物局 (ed.), *2009 Zhongguo Zhongyao Kaogu Faxian 2009中国重要考古发现 (Important Archaeological Discoveries of China in 2009)*. Beijing: Cultural Relics Press, p. 4.

Zhu, N. C. 朱乃诚 (2013). Shidai dianfeng, bingshan yijiao: Xiashiqi yuqi yipie 时代巅峰，冰山一角：夏时期玉器一瞥 (Greatest Achievement of an Age, Tip of the Iceberg: A Glimpse at the Jade Objects of the Xia Period), in Chinese Jade Culture Centre 中国玉文化中心 (ed.), *Yuhun Guopo: Yuqi, Yuwenhua, Xiadai Zhongguo Wenming Zhan 玉魂国魂：玉器·玉文化·夏代中国文明展 (The Spirit of Jade and the Soul of the Nation: Exhibition of Jade Objects, Jade Culture, and Chinese Civilisation During the Xia Period)*. Hangzhou: Zhejiang Ancient Books, pp. 58–64.

———. (2016). Xiajiadian xiaceng wenhua yuqi liuti 夏家店下层文化玉器六题 (Six Examples of Jades of the Lower Xiajiadian Culture). *Kaogu考古 (Archaeology)* (2): 95–110.

Zhuang, Y., Ding, P., French, C. (2014). Water Management and Agricultural Intensification of Rice Farming at the Late-Neolithic Site of Maoshan, Lower Yangtze River, China. *Holocene* 24: 531–545.

Index

Printed in the United States
by Baker & Taylor Publisher Services